Time-Series Python

I0062132

Contents

3

Disclaimer

Introduction

The Nature of Data

Before getting into the world of time history, let's learn a little bit about data.

Two things come to our mind when we study data. The first one is that data is a representation of the real world. Consider just one example: every day; we leave our footprint on the web with every click that we make, every like that we make and every comment that we type. All these actions say a lot about our interests, and, from this, it is possible to map our consumption profiles, topics of interest, favorite artists, films and much more. By now, you must have realized that the ads that appear on the screens of our devices are related to our taste. That is why data is a perfect representation of the real world. It can retain a domain by mapping behavior from the data.

The second core concept is that within a particular characteristic that is always present or absent, data sets can be divided into two major classes:

- Timeless data

- Temporal data

Timeless data is data that does not depend on time. For example, a dataset containing images, datasets for sentiment analysis, for character recognition, or detection of false profiles, are not temporally organized datasets.

Some datasets have an attribute of 'time,' but that is not temporal data because it is not data that gets organized over time.

Temporal data, on the other hand, is organized over time, with a 'time' attribute being an index of observation in the dataset. Cool examples of temporal data are population, and social data such as HDI, Gross Domestic Product (GDP), unemployment rate, illiteracy rate, and geographic evolution. Ongoing cases of temporal data include public, meteorological, and financial data.

What is Time-Series?

Time-Series can be defined in more than one way. The first is from a book called Computational Intelligence in Time-Series Forecasting, written by Ajoy K. Palit and Dobrivoje Popovic:

"A time-series is a time-ordered sequence of observation values for a physical or financial variable made at equally spaced time intervals Δt, represented as a set of discrete values $x1, x2, x3,...,$

etc. In practice, the sequence of values is obtained from the sensors by sampling the related continuous signals. Based on measured values and generally corrupted by noise. Time-series values usually contain a deterministic signal component and a stochastic component representing noise interference that causes statistical fluctuations around deterministic values."

That sounds complicated, and the reason it is such a detailed definition is that the book it comes from is written to address concepts and apply them in Finance and Engineering contexts.

A simpler definition of a time-series is:

"A time-series is a set of data ordered over time, in a well-defined time interval."

Much better, easier to understand.

Great! Now we need to clarify the definition.

When the second definition says, "time-series is sorted over time, in a well-defined time interval," it means that data is grouped sequentially, using two factors:

1. The recorded data of occurrence

2. The unit of time – daily, weekly, monthly, yearly, etc

The graph below shows a time-series:

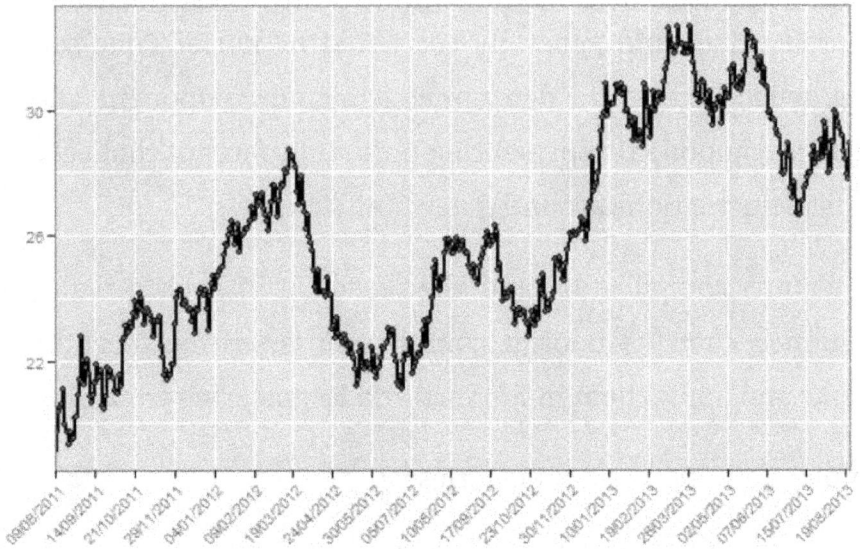

On the horizontal axis (x-axis), you can see the dates the data were recorded as blue dots. The first definition refers to these blue dots as $x1, x2, x3,..., xn$.

On the vertical axis (y-axis), you can see the attribute values for the data, in this case, $ amounts.

The last part of the first definition states that "Based on measured values and generally corrupted by noise, time-series values generally contain a deterministic signal component and a stochastic component representing noise interference that causes statistical fluctuations around the values deterministic." This has a relationship with the less trivial statistical characteristics of the time-series. These characteristics define some of the behaviors exhibited by a time-series but discussing those is outside the scope of this book.

A Brief Historical Overview of a Time Series

- Time-series theory was implemented primarily during the interwar period by Yule and Wald.
- During the 1950s, researchers at the Cowles Foundation Group developed econometric models of simultaneous or interdependent equations. These were developed to estimate the maximum likelihood methods used and were applied mostly to macro-economic models.
- During the 1960s, Holt and Winters developed the exponential smoothing technique time-series, mostly used for forecasting.
- During the 1970s, a methodology was developed to create empirical time-series models, providing a boost to applications by Box and Jenkins. This methodology states that when the information on the design is searched for, it is directed to the available data. That means it is not necessary to rely on pre-existing theories to create models.

- Box-Jenkins model forecasts were often better than large-scale macro models. However, macroeconomists found them inappropriate for a theoretical economic policy because they were empirical and thus lacked a theoretical basis.
- Until the early 1980s, there was a great deal of rivalry (see, for example, Johnston, Econometric Methods, McGraw Hill, 1986). After that came convergence.

Why is Time-Series Exciting?

No doubt, you've seen graphs showing how the stock market has evolved over the last 12 months. You've probably seen graphs showing a country's economic growth over a set period, or graphs showing the temperatures over 12 months for a country you intend to visit.

Most of you will not be aware that all of those examples are time-series because they represent numerical variables over time. In this section, we will look at what a time-series is and the important concepts that help you interpret them.

As mentioned, a time-series is a numerical variable. Time has a scale. And this scale is ordered. It can be in months, days, hours, seconds, years, decades, weeks, quarters, whatever timeframe you want to use. For example, there would be 365 values for the stock market value every day for one year. Place those values into date order, and you have a time-series.

If you opt for days, you get one value per day; if you opt to show the data by month, you get one value per month. The level of detail in your variable is dependent on the time scale you choose. Take, for example, a temperature observation for one year. You would have 12 values – one for each month. Each month would have between 28 and 31 observations, depending on how many days in the month, and that equates to 365 values per year.

The time-series resolution is equal to the number of points in a fixed period. The more points there are, the more resolution the time-series has, and that means you have more available information.

Have a look at these examples of time-series in real-life applications:

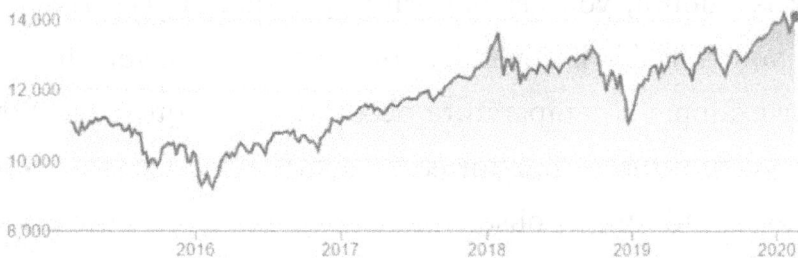

NYSE Composite
INDEXNYSEGIS: NYA

14,087.13 +48.11 (0.34%) ↑
Feb 19, 4:12 PM EST · Disclaimer

| 1 day | 5 days | 1 month | 6 months | YTD | 1 year | 5 years | Max |

Open	14,077.51	Low
High	14,115.95	

14,066.24

This is a stock market graph using the NYSE index over five years. The index shows the health of the stock market in the USA, and you can see, in the graph, the time-series trend.

Typically, mathematical techniques are used to create the trend, and, later, I will show you how this is done. For now, the trend is an excellent way to understand how your numerical variable is going up and down over time.

A country's GDP time-series is another good example. This time-series will show you the economic health of that country, and, in the example below, we can see the GDP for three major countries

– the USA, Japan, and the UK – as it has evolved over the 20th century.

Growth in Real Per Capita GDP in Japan, Britain, and the US, 1870–2008
(Natural log of per capita GDP in 1990 international Geary-Khamis dollars)

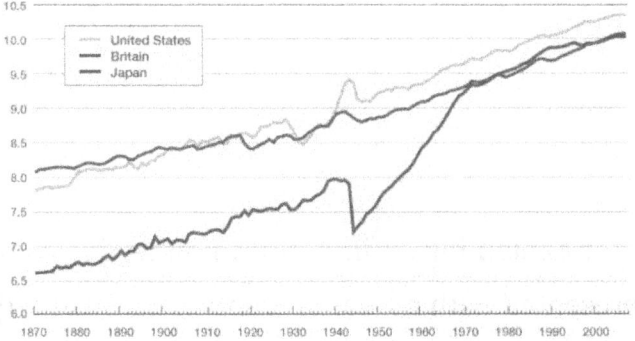

The red line shows Japan's GDP, and we can see that, following WWII, it was slow to rise until the 1970s.

In simple terms, the time-series uses historical data to show the evolution and trend over time.

The example below shows the number of visits per day to a blog. The values used are the dates between July 1 and July 31, 2019, inclusive.

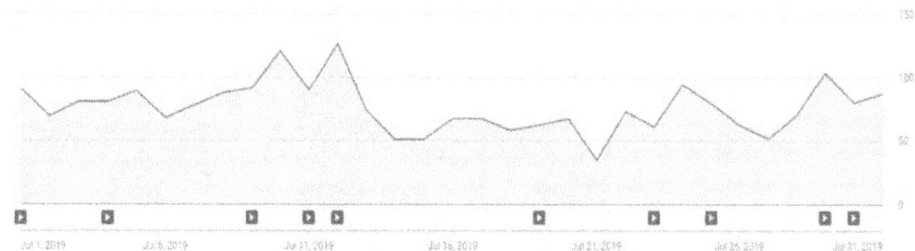

Views	Watch time (minutes)	Subscribers
2.4K	9.8K	+41
↑ 4%	↑ 39%	↑ 11%

This graph is perfect for explaining an important point. Although the time-series is represented with a line, in reality, what you see is observations per day. The points you see in the graph are the data, corresponding to the visits per day.

So, from this example, we can calculate how many visitors there were:

- For any given day in the month

- For each week in the month

- Total visitor numbers for the month.

This data allows us to predict the potential number of visitors for the next month, for a quarter, even for the entire year.

A more complex example is the time-series below, showing the annual average temperatures between the years 1200 and 2000. This is known as a "hockey stick" graph and was responsible for no small amount of controversy between politicians and scientists.

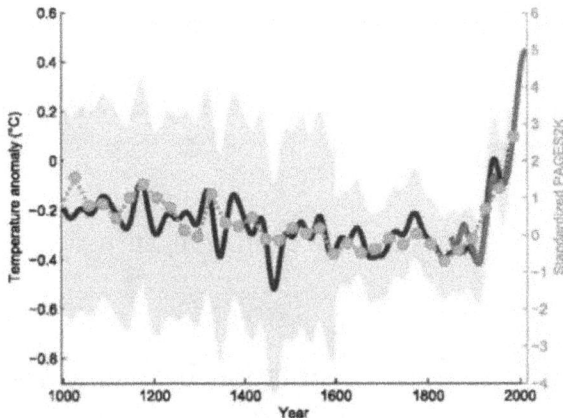

This kind of graph goes some way towards explaining the effect of greenhouse gases on global warming. The historical data was obtained from ancient trees, and the radius of the log rings used to measure the temperature.

The graph shows a light-blue shaded area; this indicates the measurement error as trees were used to measure the data and not thermometers. There is always the potential for the error to be high, but you can see that the error reduces as more historical data becomes available, and we have more accurate values to work with.

This particular graph shows two things – the exponential rise over thirty years to the year 2000 and, shown in red, a prediction until 2030.

Summary

You can see from these examples that we can glean a lot of information from a time-series:

1. We can understand the past and its evolution, intuiting the trend.

2. We can understand the current situation by seeing what has happened.

3. We can predict the future with a prediction error based on historical data.

These are the three main applications of the time-series.

Getting Started with the time-series and Its Evolution

In this chapter, we are going to warm up to the concepts involved in analyzing time series. We are going to look at 4 key elements that time series is used for with the relevant graphs and equations. We also look at 4 different ways to visualize time series.

Then we're going to look at more advanced time series concepts like autoregression models, moving average and ARMA/ARIMA models. While we introduce these concepts in this chapter, we investigate more applications in the subsequent.

What Elements of a Time-Series Are Interesting?

So far, you have a global picture of what a time-series is and what it can do for you. It is a numerical variable seen in time. You can see the values of the numerical variable in historical order. This helps you to see the evolution in numbers of this variable. And you can observe the trend in different periods according to the time scale you have.

Now I will show you the elements of a time-series with an example. I continue with the example of the blog visits presented above. Look again at the image above. There are two variables:

1. The first is the time scale, in this case, the days.
2. In the second column, the numerical variable (the one you want to study).

An observation of the numerical variable corresponds to each day. This is the time-series. You can draw the points of the time-series on a chart, type scatter plot (first chart). Or you can join the points with a line to visualize the historical trend of the variable.

Note that by joining the scatterplot points, you can create a graph where you visualize much better how the numerical variable goes up and down in time.

This is a time-series, and the graph is a time-series plot. But hold on to the idea that the data you have is what you see in the table. That is the point of the graph.

How close the points depend on the time scale you have.

Time-Series Properties

Depending on the area of study, books that talk about Time-series may present these properties in different ways and separate orders. So, for your better understanding, I decided to group the characteristics into two features: type and behavior.

Type

The type of the series is related to the structure of your dataset. Basically, time-series can be grouped into two types: the univariate series and multivariate series.

A univariate time-series contains only one attribute representing the problem domain. Just look at an example:

	Date	Temperatu
0	1981-01-01	:
1	1981-01-02	:
2	1981-01-03	:
3	1981-01-04	:
4	1981-01-05	:
5	1981-01-06	:
6	1981-01-07	:
7	1981-01-08	:

On the other hand, a multivariate time-series consists of several traits that represent a given problem domain. Have a look:

	Date	Location	MinTemp	MaxTemp	Rainfall	Evaporation	Sunshine	WindGustDir	WindGustSpeed	WindDir9am	...	Hu
0	2008-12-01	Albury	13.4	22.9	0.6	NaN	NaN	W	44.0	W	...	
1	2008-12-02	Albury	7.4	25.1	0.0	NaN	NaN	WNW	44.0	NNW	...	
2	2008-12-03	Albury	12.9	25.7	0.0	NaN	NaN	WSW	46.0	W	...	
3	2008-12-04	Albury	9.2	28.0	0.0	NaN	NaN	NE	24.0	SE	...	
4	2008-12-05	Albury	17.5	32.3	1.0	NaN	NaN	W	41.0	ENE	...	
5	2008-12-06	Albury	14.6	29.7	0.2	NaN	NaN	WNW	56.0	W	...	
6	2008-12-07	Albury	14.3	25.0	0.0	NaN	NaN	W	50.0	SW	...	

The Behavior of Time-Series

Time-series can present different behaviors according to some statistical properties that may be present and/or absent in the data. Let's introduce the characteristics that a Time-series can present:

- Stationary
- Linearity
- Trend
- Seasonality

Stationary

The stationarity of a Time-series is defined when all the values in the series vary around an average, and that average must be constant over time. Three elements characterize the stationarity of a Time-series: constant mean over time, constant variance, and the covariance between two observations x_t and x_{t-d} depend only on the distance between them, and that does not change over time.

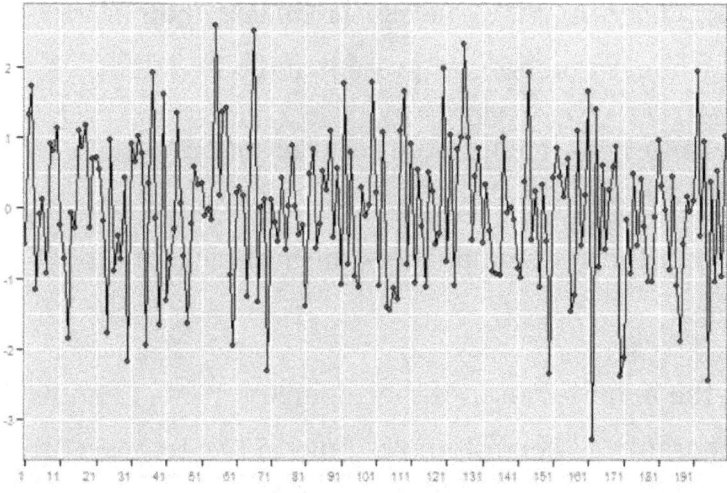

The graph of a stationary series looks like this: data varying around a constant average, which in this case is around 0 with the values between 2 and -2.

Linearity

If you already know Data Science and like it, you should know that the objective of training an algorithm is to be able to extract a model (which in the end is a mathematical function) which best represents the behavior of data to predict new values with a satisfactory degree of accuracy.

The linearity of a time-series indicates that its shape depends on its current state so that the current state can determine the model of the series. Therefore, if a series is linear, then it can be represented by linear functions of present and past values.

Since linearity allows us to check the accuracy of our measurements within the specified measuring range, linearity indicates whether the system has the same accuracy for all reference values.

For example, the manufacturer of thermometer wants to know if it provides accurate and consistent readings at five temperature levels: 210°, 208°, 206°, 204°, and 202°. Six readings are taken each time. To determine whether the thermometer is biased, individual measurements are subtracted from the reference value. The bias values for measurements with a temperature setting of 202° are calculated in the table below.

Measurement		Actual		Bias
202.7	-	202	=	0.7
202.5	-	202	=	0.5
203.2	-	202	=	1.2
203.0	-	202	=	1.0
203.1	-	202	=	1.1
203.3	-	202	=	1.3

Temperature measurements with a setup of 202° are biased positively. The thermometer indicates higher readings than the actual temperature.

For interpreting the linearity of the thermometer data, determine if the thermometer bias changes for different heat settings. If the data does not form a horizontal line in a scatter plot, there is linearity.

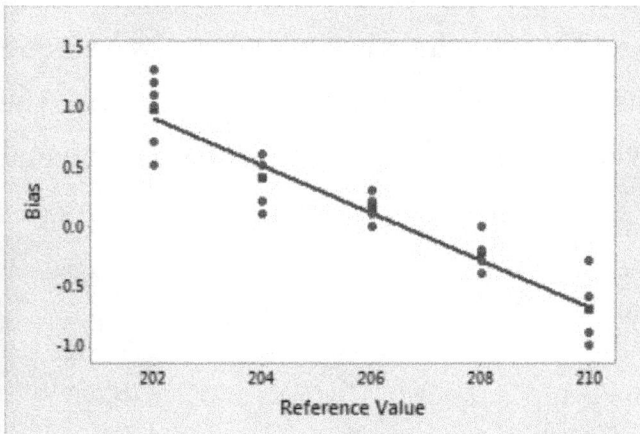

In the scatter plot, we can see that the bias changes with the increased heat setting. In lower heat settings, the temperatures are higher than the actual. In comparison, the readings of temperatures are lower than the original for the higher heat settings—the bias changes with different heat settings, which means that this data is linear.

Trend

The word trend will always be a pattern when we want to refer to something that has a given behavior and has been observed for some time; we can conclude that the trend is for that behavior to remain or change.

In Time-series, the idea of trend follows the above reasoning. When we see a graph in a time-series where the data is rising, falling, or constant and how fast these behavior changes, we see a trend. In other words, we see this behavior in the long term. The most common behaviors are the constant trend, linear trend, and quadratic trend.

Here in this graph, we can see the behavior of the indicator much more clearly. We can see that time, which is the variable under analysis here, apparently has a downward trend.

This means that, as time goes by, there is less demand for a particular product or service that is under the analysis.

Seasonality

A time-series presents some patterns of behavior. A particular pattern can be repeated at specific times over time, and that pattern is called seasonality. The word seasonality is derived from the word 'Season.'

For example: imagine a Heineken dataset for selling beer in western countries. We know that beer is a drink that is served cold and consumed in events of entertainment and the coldest seasons of the year, between December and February.

So, imagine a graph where sales have been observed for five years, where we are viewing data at annual intervals and notice that there is a peak in sales between December and February and a considerable drop in the coldest periods of the year.

Here is an example to illustrate how seasonal behavior appears in the data. Below is a graph of a seasonal series for minimum daily temperatures in Melbourne that reflects the behavior cited in the beer example:

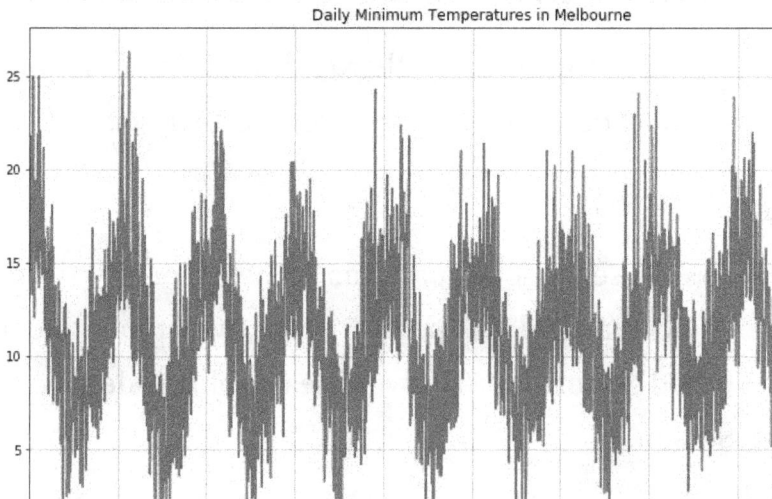

Daily Minimum Temperatures in Melbourne

The graph plots time on an annual interval, denote the seasonal behavior. Note that the series design between one year and another has a "V" shape, which shows a peak in high temperatures between the end of one year and the beginning of the other and a drop in temperature in the middle of the year. See that the seasonal pattern is repeated over the years observed, thus showing the seasonality of the series.

The Statistical Tools to Interpret a Time-Series

The first thing is to understand the time-series as a numerical variable. You can use the same tools as if it were a single numeric variable. You can use the graphics to understand the distribution. I personally use the histogram and the boxplot

- Histogram
- Boxplot
- Q-Q plot

And calculate the numerical summary

- Half
- Standard deviation
- Median
- Interquartile Range

- Maximum
- Minimum

Histogram Example

The histogram is a graph of the distribution of a set of data. It is a special type of bar graph, in which a bar is attached to the other, that is means there is no space between the bars. Each bar represents a subset of the data. A histogram shows the accumulation or tendency, variability or dispersion, and the shape of the distribution.

A histogram is a graph appropriate to represent continuous variables, although it can also be used for discrete variables. That means, using a histogram, we can graphically show the distribution of a quantitative or a numeric variable. The data must be grouped into intervals of equal size, called classes.

Let us think that water is an indispensable element for the production of a tannery; likewise, the decision of how much water to use is decisive in the finish of the leather that is achieved.

A company is defining the ideal amount of water to use in its process, for which it performs various experiments in which it collects data associated with the amount of water used and the compliance in the finishing of the finished leathers.

The tabulated data are as follows:

Intervals (Liters)	Number of finished leathers
0.0 - 1.5	3
1.5 - 3.0	15
3.0 - 4.5	18
4.5 - 6	12
6 - 7.5	7

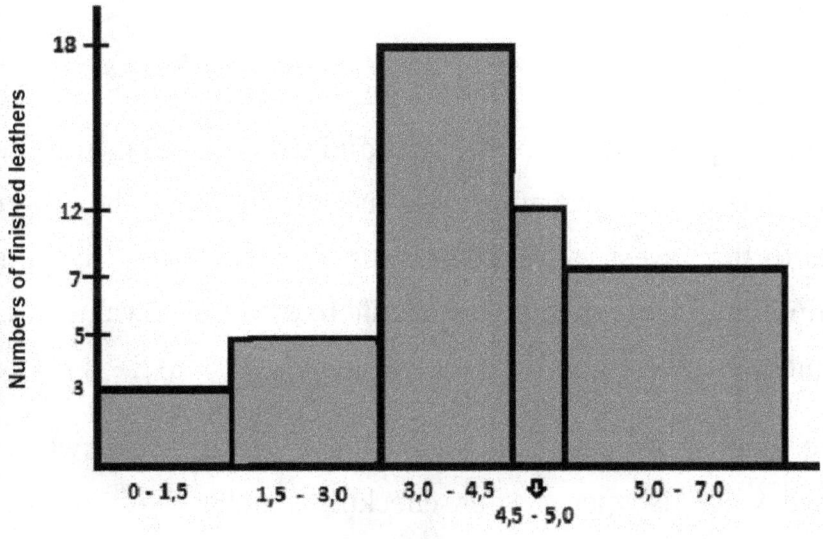

Note that the range 3.01 - 4.5 is the one with the highest number of finished leathers in good condition. In the solved example, I considered variable intervals only to illustrate a different example from those usually seen in digital and physical sources.

Example of Boxplot

Box Plot (also called box and whisker plot) is an alternative method to histogram and branch-and-leaf to represent data. The Box Plot provides information on the following characteristics of the data set: location,

- dispersion,
- asymmetry,
- tail length.

The boxplot is used to evaluate the empirical distribution of the data. The boxplot consists of the first and third quartiles and the median. The lower and upper stems extend from the lower quartile to the lowest value. The extension is not lower than the lower limit, and from the upper quartile to the highest value, it is not higher than the upper limit.

Assume that a factory produces an automotive part whose reference value is 75cm. After checking batches with out-of-specification parts, they sent two teams of workers (A and B) for training. To check the efficiency of the training, 10 pieces produced by teams A and B and 10 pieces produced by teams C and D that did not participate in the training were selected.

The data are as follows:

A		B		C		D	
75,27	74,93	74,94	74,75	75,93	73,34	75,98	76,75
75,33	74,72	75,25	74,65	76,95	74,04	75,61	76,78
74,58	74,53	75,44	74,94	75,47	75	74,2	74,74
75,01	75,32	74,62	74,92	73,6	76,18	76,44	72,58
75,71	74,05	75,35	75,46	74,85	75,33	76,84	72,86

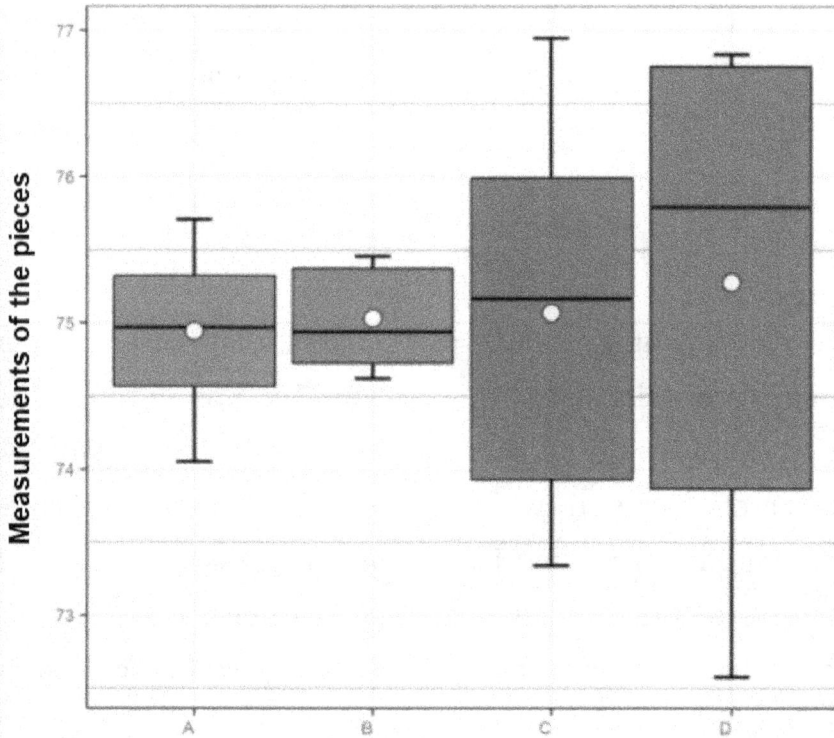

BoxPlot By Group

Analyzing the graph, we can see that:

1. Teams A and B produce parts with less variability, indicating that the training had the desired effect;

2. Team D is the one that produces parts with greater variability;

3. Team B is the one that produces parts with the least variability.

Considerations: As the parts of teams A and B have less variability and with an average value close to the reference value, it is worth sending the other teams for training.

Q-Q Plot Example

The QQ plot (quantile-quantile plot) is used to examine whether a certain distribution is normal or not. For this purpose, the quantiles of the empirical distribution (measured values of the sample) are compared with the quantiles of the standard distribution in a graphic. The number of possible quantiles corresponds to the number of measured values. If the data is (approximately) normally distributed, this represents the results in a diagonal line.

Assume that the height in cm is measured for a sample of 10 men: 180, 150, 160, 170, 190, 200, 210, 180, 170, and 190.

The QQ plot is to be plotted to assess whether there is a normal distribution. To do this, the empirical quantiles on the y-axis and the theoretical quantiles of the standard normal distribution on the x-axis are plotted:

Normal Q–Q Plot (series Y)

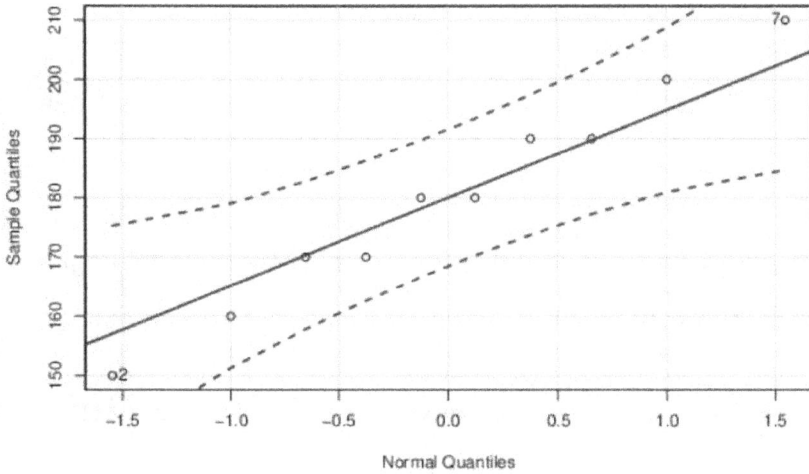

In the plot, the approximate diagonal on which the points are located indicates a normal distribution of the data.

Now let's move on to the next section to delve deeper into the time-series.

We have already seen that a time-series is a set of data or observations that refers to one or more variables, and that is ordered chronologically. Time-series are very important in economics because almost all the variables are collected over time. That means it is interesting to see the evolution of a variable over time, not the specific value at a given moment.

Hence, whenever economic variables are analyzed, economic cycles or trends are discussed.

Since the order of the data has vital importance, one must bear in mind that this modifies the analysis and interpretation of the data. Therefore, econometrics, which is responsible for finding and estimating relationships between economic variables, should take this fact into account.

Time-Series Analysis Based on its Definition and Characteristics

Given that the order of the data matters, we can say that the observations are not independent. This means that the past can affect the future. Econometrics should be aware of this characteristic and use mathematical tools that allow it to make estimations reliably. Definitely:

1. The order of the data matters.
2. The observations are not independent.
3. When estimating relationships, keep in mind that they are not independent.

Therefore, you must use different mathematical techniques and statistics.

Knowing this, then one may ask:

- What exactly does it mean that the observations are not independent?
- With what techniques are the time-series data analyzed?

Temporary Dependency

The answer to the first question refers to temporal dependence. A variable has a temporal relationship when past data affects the value of the variable in the future. For example, the long-term global GDP has a prolonged upward trend. This means that economic growth is sustained over time; therefore, what happened in the past affects the future.

On the contrary, if we roll a dice and note the date on which we roll it, we will see that there is no relationship between past and present data. In the latter case, the past does not affect the future.

Techniques for Analyzing Time-Series Data

There are many techniques to analyze time-series data. However, what is usually easier is to use a regression model. Of course, a regression model takes into account the type of time-series with which it works.

One of the most used and simplest techniques could be to modify the series or take it into account in the model. For example, eliminate the trend of a series of GDP or include a trend variable in the model. Although it is not the subject of this definition, I will give a very simple example so that you can understand it.

Let's look at the following graph:

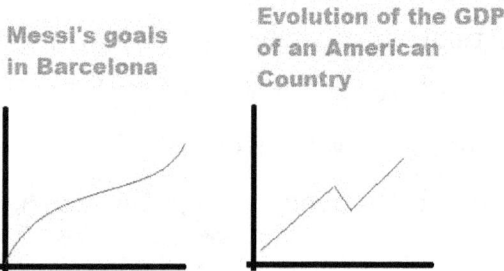

Messi's goals in Barcelona

Evolution of the GDP of an American Country

If we calculate a regression model of the two previous series, the calculations indicate that there is a statistical relationship. However, the goals that Messi marks with the growth of an American country have nothing to do with it. However, by eliminating the trend component, it would be inferred that they have no relationship.

What is described in the previous paragraph is something that happens many times with series that are related, but when the research is done well, they do not have this problem.

A Broader Explanation of Time-Series Characteristics

Time-series analysis is an important instrument in understanding the market and in formulating action plans and strategies. The history of a variable can be used to identify periods of growth/decrease, seasonality, and also to "predict" future observations.

Time-series models are widely used to assess the behavior of a variable over time. An ice cream shop, for example, tends to sell more ice cream during the summer, while the market for coffee shops tends to be more heated in the winter.

Would it be possible to "forecast" the demand for ice cream for the summer of 2020?

In fact, the statistical models for time-series use the variable's historical past to project future observations. Thus, one can have an idea on average how the variable will behave in the coming periods.

Time-Series: Autocorrelation

The most well-known and used statistical models, such as Linear Regression and Generalized linear model (GLM), are suitable for modeling variables in which observations are independent.

The generalized linear model expands the linear model in such a way that the dependent variable is linearly linked to the coefficients and covariates with a special feedback function. This model also allows the dependent variable to have a non-standard distribution. The generalized linear model includes the most commonly used statistical models, like linear regression for responses that are normally distributed.

In a time-series, there is no way to disregard the dependence structure of the observations.

For example, the quantity of ice cream sold in February may be related to the quantity sold in January, which in turn may be related to that of December and so on. Thus, the use of these models can generate biased results that do not reflect the reality.

Autocorrelation is defined as an observation at a given moment that is related to past observations.

Observations can be autocorrelated in several orders. The first-order autocorrelation characterizes series where an observation is correlated with the immediately previous observation (February and January, for example). Second-order autocorrelation describes a time-series where an observation is correlated with observations at two different time units in the past (February and December, for example).

The identification of the autocorrelation is made through the Autocorrelation Function (ACF), as shown below. In addition, tests such as Durbin Watson's help to identify the first-order autocorrelation.

In the ACF example shown below, the significance of the autocorrelation order is assessed using the confidence intervals (in blue). Thus, this series has the first-order autocorrelation, since the point at lag 1 is significant.

The numbers in the graph below are correlation coefficient. So, they are each correlated with the previous observation. A correlation coefficient is a number between 1 and -1 that respectively defines a positive or negative correlation. A value of zero indicates no correlation.

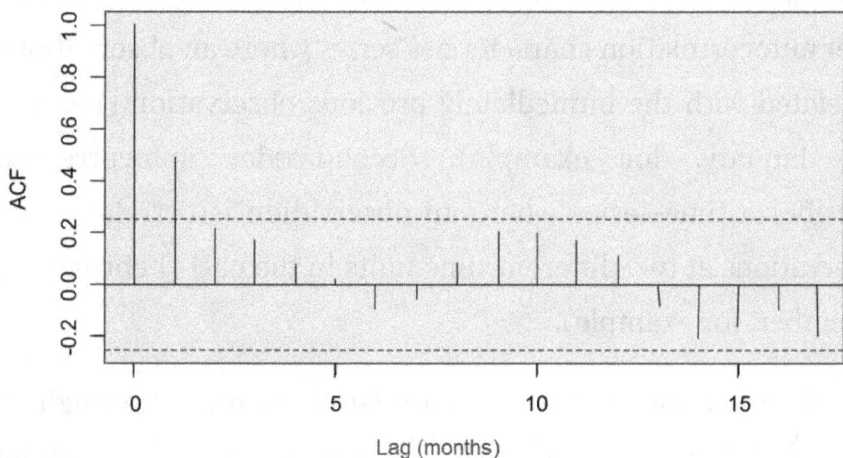

Lag (months)

Time-Series: Trend

The trend of a time-series is defined as a pattern of growth/description of the variable over a certain period of time.

There are specific tests for identifying the trend, such as the Wald Test and the Cox-Stuart Test. However, a widely used technique is the adjustment of a Simple Linear Regression to determine the slope of the trend line.

It is worth remembering that the adjustment of the Linear Regression, in this case, can lead to biased results. And to avoid this problem, robust estimators for the autocorrelation can be used. This topic will be discussed in detail in the next chapters of this book.

In the example below, we consider the data referring to the number of international passengers in air transport between the years 1949 and 1961. A clear growing trend can be seen in the series over the years.

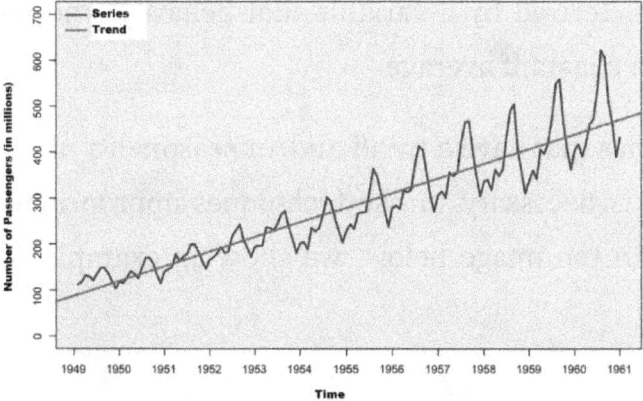

Time-Series: Seasonality

Seasonality can be defined as patterns of behavior that are repeated at specific times of the year. For example, the number of passengers using air transport is generally higher during school vacation periods than in other months of the year, a fact that can be seen in the image presented above.

Seasonality can often be identified visually. One of the formal techniques used is the identification of significant autocorrelations of a given order. For example, if there is an

43

annual seasonality in a series, the autocorrelation of order 12 is generally substantial.

Time-Series: Stationarity

Stationarity is an important concept in the modeling of time-series and is characterized by a variable that behaves randomly over time around a constant average.

Basically, time-series that have a trend and/or seasonality are not stationary, and it is necessary to use techniques appropriate to such a situation. In the image below, we show an example of a stationary series.

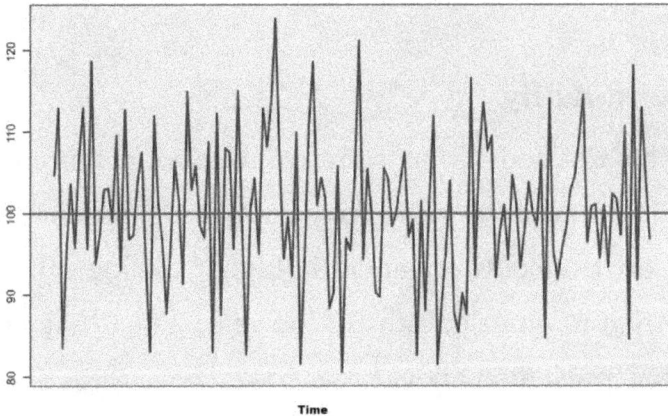

After understanding the behavior of the series and identifying the elements, it is possible to verify which model is most appropriate.

To Recap

What we have learned here:

- Nature of the data (temporal and timeless data)
- Definition of time-series
- Time-series properties and characteristics (type and behavior).

Applications of Time-Series in Making Production Forecasts

Generation production forecasts are an everyday activity of the petroleum engineer who has been developing using the implementation of computational tools based on the models' decline curve and type curves. These models have shortcomings in the accuracy of forecasts, mainly for two reasons: first, it has the excellent condition that the wells are operating in a pseudo-stable state, and second, adjust the behavior of production to a trendline, which is extrapolated in time for the prognostications.

In this section, as an alternative to these models, the use of time-series forecast generation has been proposed because they include both the trend and cyclic components and seasonal production data. Comparing the absolute error exists between the actual data and forecasts obtained by conventional methods and the application of time-series models. Using these series yielded a better fit historical data, evidence that other trends may occur in decline (as the cube, for example) and increase the accuracy of forecasts generated.

Currently, the techniques used in generating production forecasts are based on models of "decline curve" (Arps, 1945) and "Curves Type" (Fetkovich, 1980; Palacio & Blasingame, 1993), whose use is reflected in the emergence of commercial software for tracking and forecasting production wells.

These programs integrated analysis techniques pose data, aimed at obtaining production forecasts, from generation and extrapolation of trend lines. However, analyzing these techniques can identify some points considered as disadvantageous:

- Require producing wells to be in a pseudo-stable state, that is, the pressure of flowing downhole constant. This condition is hardly satisfied due to various situations (gas accumulation in the annulus of the well, changes in the efficiency of artificial lift systems, formation damage, reduced capacity of the pipes, etc.) generate variations in the pressure conditions of the well.

- They assume that the variables have a parametric behavior, i.e., governed by normal trends that can be extrapolated. Actually, several things can happen here; for example, a well has several production trends throughout its life. As the phenomena mentioned above occurred, the data indicated fluctuations around the trend line, or extrapolating this trend does not allow us to obtain a good fit for the forecast.

- They have full information failures when it lacks sufficient and necessary data to implement them. This being a strong limiting field was not always updated data in areas such as fluids and petrophysical analysis of pressures.

- Do not involve the probability of occurrence of certain circumstances (such as faults in production systems, interventions wells, changing operational conditions) responses from the production system to the actions taken.

Given these reasons, the approach aims to seek new methodologies for monitoring the performance of a reservoir and predicting their future behavior, from the application of time-series.

These series are based on the assumption that current data can be calculated from the analysis of the above data, but not simply considering the tendency of the same, but include the fluctuations in variables over time, represented in both cyclical and seasonal components.

The aim is to assess how much can be improved predictions made by the time-series with respect to the results obtained with traditional models.

Decline Curve

Since the first half of the twentieth century, the oil industry has used the decline curve as the main tool for obtaining production forecasts, even though some of its considerations hardly met in practice.

A decline curve has two parameters that describe it. First, the rate of decline (D), which reflects the rate at which decreases the production of a well in terms of average daily, monthly or yearly, and is calculated using Equation 1, where q_1 and q_2 are respectively production rates of the well measures at times t_1 and t_2.

$$D = \frac{(q_1 - q_2)}{q_1(t_2 - t_1)}$$

And in the second instance, the exponent of declination (n), that is reflected in the trend of decline curve (see Figure above), which shows that this may vary from one situation in which the decline is severe (declination is exponential with an exponent n = 0), to a condition of slow reduction of the production rate (harmonic decrease, with an exponent n = 1) through intermediate cases (hyperbolic decline, 0 <n <1).

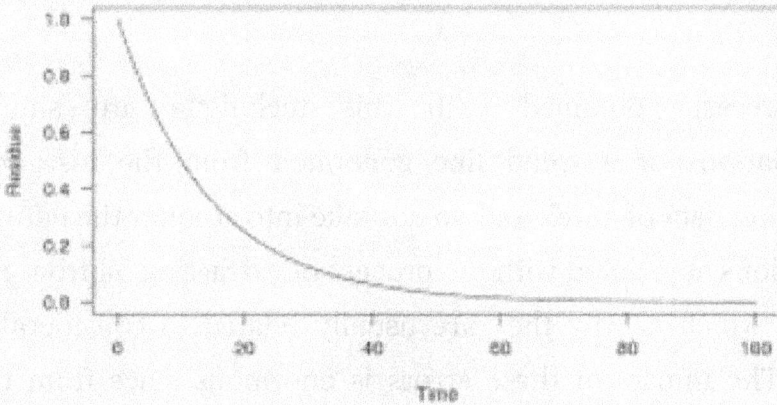

The exponent declination can be determined in different ways, either by generating graphics, identifying trends, or using curves type.

Once we determine the declined exponent, then we can use the model proposed by Arps (https://petrowiki.org/Production_forecasting_decline_curve_analysis) (presented below) to calculate future rates of production. The produced oil accumulated in an instant determines the life of the well or the time remaining to reach a certain production rate.

$$\frac{D}{D_i} = \left(\frac{q}{q_i}\right)^n$$

The forecasts obtained with this technique are simple extrapolations of a trend line generated from the historical production data of a well and do not take into account the natural fluctuations associated with the process of extracting oil from the ground, which is why they are usually related to considerable errors. The impact of these errors is enormous since from the production forecasts, many of the investment decisions are made in the production fields, operational expenses are quantified, production commitments are generated, and oil sale contracts are generated, among many other aspects related to the management of an oil asset.

Advanced Time-Series Concepts

As already mentioned, a time-series is "a sequence of empirical data ordered in relation to time; that means, it can be plotted against time," so it is applicable to a set of data recorded periodically, whether a daily, weekly, half-yearly, monthly or annual record. Among others, some practical examples are the total annual sales of warehouses, the quarterly value of the GDP, or, in this case, the production of one or more oil wells.

To describe a time-series, the following components must be identified:

a) *Trend or trend component:* Is the result of factors affecting the data in the long term, which generate a gradual and consistent pattern of variations in the same series. In the case of a well, the trend of production data is usually the decline due to phenomena such as depressurization of the reservoir, depletion of fluids present in it, or the occurrence of formation damage. According to traditional methodologies, the trend may be exponential, hyperbolic, or harmonic. But the other trends, such as linear, quadratic, or cubic, may be considered in data analysis.

b) **Cyclical and seasonal components:** Both cyclical and seasonal variations can be understood as the undulating movement of data above and below the trend line. But the difference between one and the other has to do with the range of data being analyzed. In the case of cyclical components, it refers to processes that occur more spaced out within the observation interval while seasonal components occur with less separation in time.

For example, when talking about the operation of an oil well, a seasonal component may refer to the variations in the production rate that happen every day (in the hottest hours, the rate increases, while in the coldest nights, it decreases). In contrast, a cyclical component may be given by processes of gradual reduction of the flow rate generated by the damage to the formation, and a subsequent increase in the rate, after performing periodic stimulations to the well.

c) **Irregular component:** which relates to data that are far off-trend, such as when a production stop occurs in a well. In this case, it is necessary to make decisions regarding the handling of these abnormal values, since including them in the calculations it would add a large amount of noise to the description of the abnormal values. For this reason, a value is usually estimated to replace the anomalous data. The time-series models can be classified into four large categories, which can be identified by applying the Box-Jenkins Methodology by which the simple autocorrelation (S.A.F.) and partial autocorrelation (P.A.F.) functions are calculated. The typical forms of these two functions are presented in the Figure below.

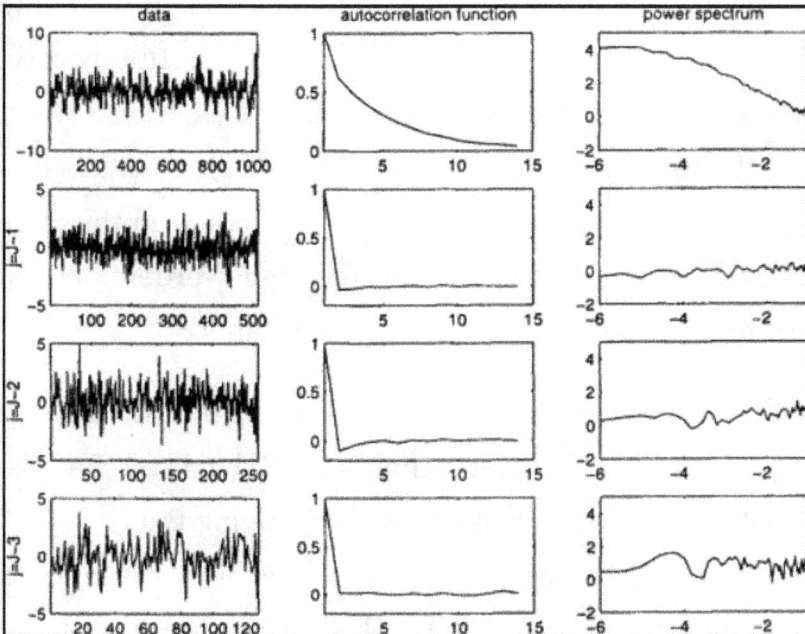

Autoregression Models

One of the characteristics that a time-series can have is that the values of the study variable (in this case, production) depend on its previous values. In mathematical terms, an autoregressive (AR) model calculates the value of the variable and at instant t using the equation shown below.

$$y_t = a + by_{t-1} + \epsilon_t$$

Where 'a' and 'b' are time-series constants, and 'ϵ_t' is the white noise assumption. y_{t-1} is the value of y at instant t-1. In this case, an AR model of order one is being referred to, because when calculating the value of the variable of interest (y) at instant t, is taking a previous data to which you want to predict (in this case y_{t-1}). Similarly, there are expressions for higher-order models, including a higher number of adjustment constants.

Additionally, the determination of the order of the model (i.e., the amount of previous values that should be used for the calculation of future value) is done by determining the "P-value" that is considered a measurement factor of how likely the model's predictions will come close to reality. This parameter ranges from zero to one, and the higher the "P-value," the more likely the proposed model will describe the data accurately.

Moving Average Models

In some cases, the data of a time-series depends on the white noise that is associated with each of the observations that make up the series. In this case, a white noises error term is added (ε_t). In addition, a second order white noise term is added that is related using a paramer (θ_i). The general form of a moving average model is presented in the equation below.

$$y_t = a + \epsilon_t + \theta_1 \epsilon_{t-1} + \dots + \theta_q \epsilon_{t-q}$$

Where a is the mean of the series, $\theta_1, \theta_2, \theta_q$, are parameters to be determined for the model, and q is the model order, i.e., the number of past values affecting the calculation of the current value. ε_t is a white noise error term at time t

If the model order moving average is 1, their expression would be presented in the equation below.

$$y_t = a + \epsilon_t + \theta_1 \epsilon_{t-1}$$

"ARMA" and "ARIMA" Models

The analysis of RA and MA models are simple, as one is simply considering a component of the time-series (the trend of the latest data or the error of the latest data, respectively). Still, it is unusual for this to happen, as in most cases, both influences coexist.

In these situations, we must resort to autoregressive models of the moving average, better known as ARMA(p,q) models, where p and q respectfully represent the autoregression order and the moving average order. The general expression of an ARMA(1,1) model would be the same as presented in the below equation.

$$y_t = a + by_{t-1} + \epsilon_t + \theta\epsilon_{t-1}$$

The models presented (RA, MA, and ARMA) have a common characteristic: they are used to represent stationary stochastic processes. That means the variables are ordered chronologically, and the probability distribution of each observation remains constant over time.

However, in some cases, this condition of a stationary time-series is not met, and making forecasts using models such as those already seen will have some associated error. For this reason, Box and Jenkins proposed ARIMA models (https://www.machinelearningplus.com/time-series/arima-model-time-series-forecasting-python/), which, in addition to containing an autoregressive and a moving average component, include an integration component. Hence its name stands for **"autoregressive integrated moving average."**

This term of integration numerically intended to lead a non-stationary series to have a stationary behavior. In this case, the numerical models are more complex, as presented in the following equation.

$$y_t = a + b_1 y_{t-1} + \dots + b_q y_{t-q} + \epsilon_t$$
$$+ \theta_1 \epsilon_{t-1} + \dots + \theta_q \epsilon_{t-q}$$

Precisely because of the large amount of data to be analyzed, the complexity of the calculations and the huge number of options for orders of the models is concerned in the process, it is recommended that this type of statistical analysis must be carried out by personnel with great experience in handling data with the help of specialized software.

Methodology Analysis of Production by Time-Series

To assess the accuracy of the predictions obtained using time-series analysis of production data of three wells located in the basin of the Eastern Plains was performed using the discussed methodology and comparing these results with which they were obtained employing four commercial programs:

- Oilfield Manager (OF M)
- ValNav
- Topaze
- Ryder Scott Forecast

The method used to predict the production employing time-series is summarized in four steps:

1. Handling of irregular data,
2. Historical trend adjustment,
3. Preparing the time-series model
4. And forecasts generation.

1. Handling of Irregular Data

Including the zero production values in the analysis of the data is not convenient. Let's assume that a failure had occurred that prevented the production of a certain volume of oil on a specific day. If this failure had not occurred, the production of the well would not have been zero. So, it's not accurate to include it in the oil production capacity calculations. Therefore, it is recommended to assign some value to the date in which the variable data appear. In this case, the "moving average" is used, a technique in which the missing value is replaced by the average of the data from the days before the date in question. In this case, data from the week prior to the event were used, according to the following equation.

$$q_i = \frac{q_{i-7} + q_{i-6} + \dots + q_{i-2} + q_{i-1}}{7}$$

This assumption is based on a statistical principle, according to which, for a data group with normal distribution, the expectation is equivalent to the average.

2. Historical Trend Adjustment

Regression analyses are statistical processes that seek to establish the relationship between variables. If we have measurements of a certain variable (for example, the production of a well) over time, we will try to establish the behavior of that variable with respect to time using an equation.

In this case, the application of the least squares method proceeds to verify the adjustment of production data to different trends: linear, quadratic, cubic, and exponential.

The below table presents a summary of the information criteria that allow the adjustment of the data to the proposed trends to be verified:

Model	Linear	Quadratic	Cubic
R^2 Adjusted	0.7086	0.85	0.86
MSE	$4,2 \times 10^6$	$2,2 \times 10^6$	$2,0 \times 10^6$
AIC	15.25050	14.61050	14.52807

- **_The determination coefficient (R2):_** It represents the relationship between the variations (deviations) explained by the model and the existing total variations. That means a value of R2 equal to one means that the model describes all the values that present some variation with respect to the trend. As a decision criterion, the model that shows the R^2 value closest to the unit must be selected.

- **The MSE:** This criterion, as its name indicates, measures the average of the errors of each data squared. Therefore, the smaller the MSE value, the better the model studied.

- **The Akaike information criterion (AIC):** This is a measure of the relative quality of the model being considered for the description of the phenomenon. It presents an estimate of the information lost when the model being considered is used to describe the data. In other words, the smaller the value of the AIC, the less information is being lost, and therefore, the model is more accurate.

- **The Bayes Information Criterion (BIC):** Analogous to the AIC, this criterion allows the best model to be identified. The BIC relates to the deviation of the real data from the model, including some Bayesian corrections. The definition says that "BIC includes the unexplained variation of the dependent variable," which indicates the weaker the description of the variable, the higher the value of the BIC. In other words, the best model is the one with the lowest value of the BIC.

As we can see in this case, the best adjustment was presented for the cubic trend, to the detriment of the exponential adjustment that is commonly used.

The Figure below shows the historical production observations of one of the study wells, along with the proposed trend curves.

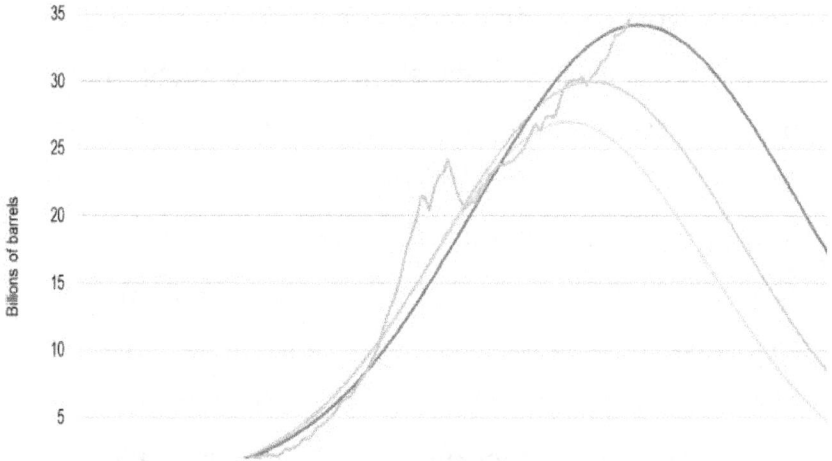

3. Preparing the Time-Series Model

Two very important aspects must be determined at this point, the type of model and the order of the model.

The first aspect is related to the identification of an equation that more precisely allows describing the historical production data, including its trend and its cyclical and stationary components. To this end, the BoxJenkins methodology must be applied, with which the "Simple Autocorrelation Function" (S.A.F.) and the "Partial Autocorrelation Function" (P.A.F.) of the data are generated. The Figure below shows the graphs obtained numerically for the data of the study well, verifying that an ARIMA model can be used in the subsequent analysis since the graphs coincide with that proposed in the figure of *Irregular Components* part of this book.

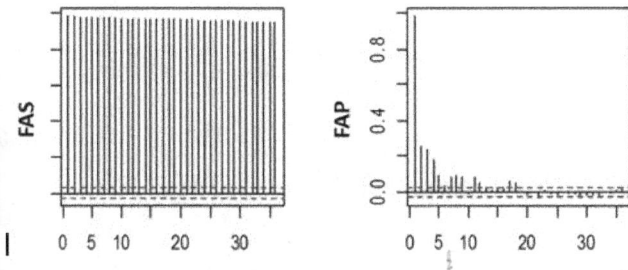

Then, through trial and error, it is necessary to determine the order of the model, i.e., how much backward data should be used to calculate a new value in the time-series.

This analysis is performed by calculating the "P Value," which relates to the adjustment of the model, as mentioned earlier. As shown in the table below, in this case, the selected model is an ARIMA (10, 1, and 10) because it has the highest p-value of 0.9965.

Order Model	1,1,1	2,1,2	3.1.3	4,1,4	5,:
Value P	$8,9\times10^{-10}$	$5,07\times10^{-9}$	$4,01\times10^{-9}$	0.01109	0.0:
Order Model	7,1,7	8,1,8	9,1,9	10,1,10	11,:
Value P	0.07178	0.0035	0042	0.0065	80

4. Forecasts Generation

This is the final stage of the process where calculations are made based on the obtained model to establish future values.

Due to the order of the model, the cyclic and stationary components affect the forecasts in the short term (as can be seen in the figure below). Still, this effect is reduced when it is intended to continue advancing in time, getting to obtain a trend line after two months of forecasting simply. In this case, the production was predicted from day 5006 of operation of the well. It should be noted that between days 5045 and 5170, the production stops were recorded that the model developed is not able to predict.

The next figure shows a comparison of the results obtained for the four commercial software and the forecasts generated with the time-series model.

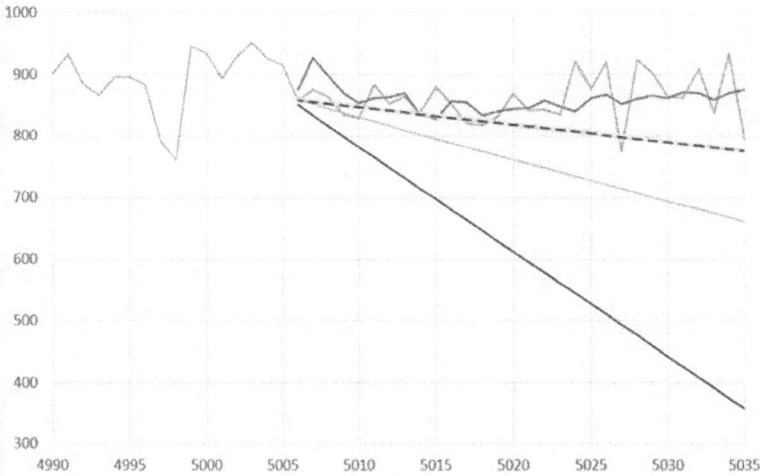

Results

The table below shows a summary of the errors obtained after generating production forecasts by five methods for a period of one month for three different wells. Since some programs use similar algorithms, the results obtained from them can be very similar.

Technique	Absolute error (%)		
WATER WELL 1 2 3			
OFM	5.34	7.18	1
Topaze	28.58	15.51	2
ValNav	11.64	9.41	2
Forecast	5.39	8.31	1

In this case, the data generated by the Oilfield Manager (OFM) and Forecast programs have very close values, which is why they are superimposed lines graphically, and the error percentages of both programs are practically the same.

As it can be seen, the implementation of the time-series methodology allows for a significant reduction in prediction error, thanks to the fact that these models consider both the production trend of the well (which is also implicit in the software tools) and the oscillations that naturally occur in the well.

The analysis was repeated for a more extended period (one year), obtaining the results presented in the next Table. The accuracy of the results is inferior, because in a period as long as one year, the small initial deviations of the models become large margins of error.

Technique	Absolute error (%)		
WATER WELL	1	2	
OFM	75.20	55.77	
Topaze	257.78	62.37	!
ValNav	61.73	72.40	(
Forecast	73.85	51.23	⸴

Although the behavior of the time-series model approaches a trend line after a few observations, there is less error compared to the results of the different programs.

From the analysis of the production data, it can be seen that the exponential adjustment is not always the one that allows the best description of the data behavior.

Although production forecasts generated by a time-series model reduce their effectiveness in the long term, it continues to be a tool with greater precision than traditional methodologies.

It is necessary to define a calculation algorithm for each well since the generalization of historical analysis and prediction techniques leads to the occurrence of significant errors.

Basics Concepts of Time-Series Analysis With Python

This chapter is all about getting started with Python in Time Series. We start off with why Python is the best software for analyzing time series. Then, we go through a step-by-step procedure to install Python on your computer.

Then we go through how to read and handle Time Series data into Python.

Why Use Python for Time-Series Data?

Python is an open and free general and dynamic purpose programming language. It is considered the easiest language to be read and learned. It has wide versatility, which is why it can be used in both the academic and business fields. It has object-oriented programming and a powerful computational capacity through the use of algorithms written in Fortran or in C. It also has a huge catalog of modules for all kinds of purposes: database management, graphics, artificial intelligence, etc. In recent years it has stood out for its use in the field of data analysis.

The reasons the data scientists think are worth highlighting Python as the best program for Time-series analysis are:

Python is Great for Beginners

First, let's answer a simple question: what makes a programming language "easy to learn"?

In my view, two factors are essential:

- Simplicity
- Learning speed

Simplicity

Python has several features that make it a so-called simple language.

First, it is a high-level language, that is, it is not necessary (but advisable) to know hardware details, or low-level languages to start learning Python.

Requires less code to complete basic tasks (such as loops, decision structures, exception handling, class declaration, etc.) when compared to other languages, such as Java (code 3 to 5 times smaller) and C (5 to 10 times smaller). Smaller code = less chance of error!

Easy Learning Curve

Because it is a simple programming language, anyone who starts programming in Python will quickly be developing sophisticated and robust programs!

By learning the basics of Python, you will be able to develop:

- Web systems with Django, Flask, Pyramid, among other web frameworks;
- Multiplatform applications with Kivy (learn more about Kivy by clicking here);
  - Link: kivy.org/
- PyGame games;
- Graphical interfaces with Tkinter or PyQT;

Versatility

Python has more than 130,000 third party libraries (exactly 135042 - at the time of writing)!

These libraries make Python very useful for specific purposes, from traditional web development, or image processing, to cutting edge technology, such as artificial intelligence, Machine Learning, and Deep Learning.

The language that helps a biomedical doctor to study the genetic sequencing of a species can also be the language that allows a data scientist to identify fraud and money laundering attempts worldwide!

Multiplatform

Because Python is an interpreted language, not compiled for a machine language, it can run on different platforms.

But first, let's understand the difference "interpreted vs. compiled"!

An interpreted language is one in which its source code is read by an interpreter and converted into executable code, which will be executed by a virtual machine. In Python, this process is divided into four steps:

- *Lexical analysis:* Breaking the source code into tokens (strings with identified meaning).
- *Parsing:* Process of taking these tokens and generating structures that show the relationship between each symbol (in the case of Python, this structure is an Abstract Syntax Tree - AST).

- **Compilation:** Transformation of an AST into one or more Code Objects ("pieces" of executable code).
- **Interpretation:** In possession of the Code Objects, the Interpreter then executes the codes represented there.

A compiled language is processed by a compiler directly from the source code for machine language specific to a processor and operating system.

Therefore, Python (which is an interpreted language) runs on different platforms, as it is enough to have an interpreter to process the source code on any system or processor. And because of this, there is no language like Python that is so intuitive for Time-series analysis for scientific or other types of research work that requires a massive amount of data visualization.

We shall start off with the basics of installing Python on your computer system.

In addition, Python is a modern programming language that can be legally obtained from the Internet for free. This is perhaps one of its main advantages it offers since scientists and their students can easily carry out their research without worrying about situations such as the acquisition of software licenses. Furthermore, the Python language lends itself naturally to applications in science, since several teams have contributed open-source libraries in various fields of knowledge, which extend the capabilities of this programming language.

How to Install Python On Your Windows / Linux / MacOS System

Python is an interpreted programming language with a multi-paradigm which supports object-oriented programming, imperative programming, and functional programming. Because of its simple and very readable syntax, it is a perfect option for those who want to learn to program. And since it comes with multi-paradigm and multiplatform features along with many modules and functions, it is also an excellent option for experienced developers too.

Whether you are a beginner or an expert, to use Python on your system, you need it installed on your computer. Therefore, you must follow the below steps to install (or update) Python as per the operating system of your computer.

Windows

Microsoft Windows is the most commercial and most commonly used desktop operating system in the world. Generally, when you buy a new computer, it is a default version of Windows. Therefore, the different ways of installing Python on these systems will be detailed below.

Python.org is the official site of the programming language. There, you will find all the information regarding the language, its documentation, and the links to download the installers of the different versions of Python that are found with the "Downloads" button.

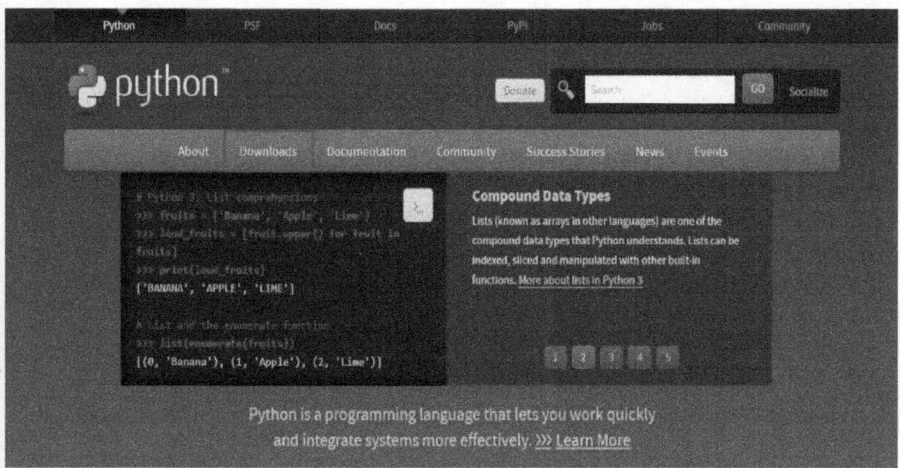

If you point the cursor on the Downloads button, a more detailed menu on possible download options will be displayed.

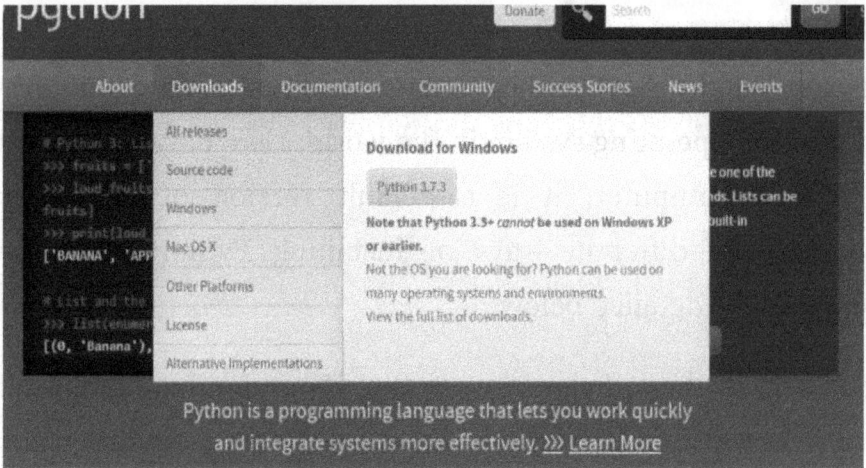

You can download the installer by pressing the "Python 3.7.3" button or, in case you need to download another version, you can press the "Windows" button. When doing so, it loads a screen similar to the following.

Python >>> Downloads >>> Windows

Python Releases for Windows

- Latest Python 3 Release - Python 3.7.3
- Latest Python 2 Release - Python 2.7.16

Stable Releases

- Python 3.6.9 - July 2, 2019
  Note that Python 3.6.9 cannot be used on Windows XP or earlier.

  - No files for this release.
- Python 3.7.3 - March 25, 2019
  Note that Python 3.7.3 cannot be used on Windows XP or earlier.

  - Download Windows help file
  - Download Windows x86-64 embeddable zip file
  - Download Windows x86-64 executable installer
  - Download Windows x86-64 web-based installer

Pre-releases

- Python 3.6.9rc1 - June 18, 2019
  - No files for this release.
- Python 3.7.4rc1 - June 18, 2019
  - Download Windows help file
  - Download Windows x86-64 embeddable zip file
  - Download Windows x86-64 executable installer
  - Download Windows x86-64 web-based installer
  - Download Windows x86 embeddable zip file
  - Download Windows x86 executable installer
  - Download Windows x86 web-based installer

Choose the desired Python version from this page, here I am downloading the version for 32 bit so be sure about the bit of your OS to get the same version of python. Once the package has been downloaded, you can install it by double-clicking the installer, and it will open a screen similar to the following:

Check that the box that says, "Add Python 3.7/your downloaded version to the PATH" is active and then press "Install Now." It will load a screen with a progress bar that will indicate each step that the installation process does.

Once the progress bar is full, a screen with options to start a Python tutorial will appear here, consult the documentation or see the changes that are made with respect to previous versions in the version that was installed is displayed. Then press the "Close" button to complete the installation process in the system.

Microsoft Store: Python can also be installed from the Microsoft Store. "Python" is added in the Microsoft store's search engine; from there, select Python.

Once loaded, press the "Get" button to get started with the download process. When the download is finished, Python will be installed on Windows.

Linux

Linux distributions are the operating systems preferred by programmers and software developers as it is the most suitable environment for their activity. These systems have a highly versatile command interpreter, and adaptable to the needs of each user as well as have highly customizable graphic interfaces in the same way.

Fortunately, Linux distributions come with Python installed by default, but sometimes the version of Python that is installed on the system is not what the user wants or needs. Therefore, shortly

I will discuss the steps to follow to install the required Python version below in case it is not installed on the system.

The first step is to access the official Python site, Python.org, where the files to be downloaded will be found. Similar to the Windows process, you need to position the mouse cursor on the "Downloads" button after that a menu will be displayed containing options to choose the system in which Python will be installed.

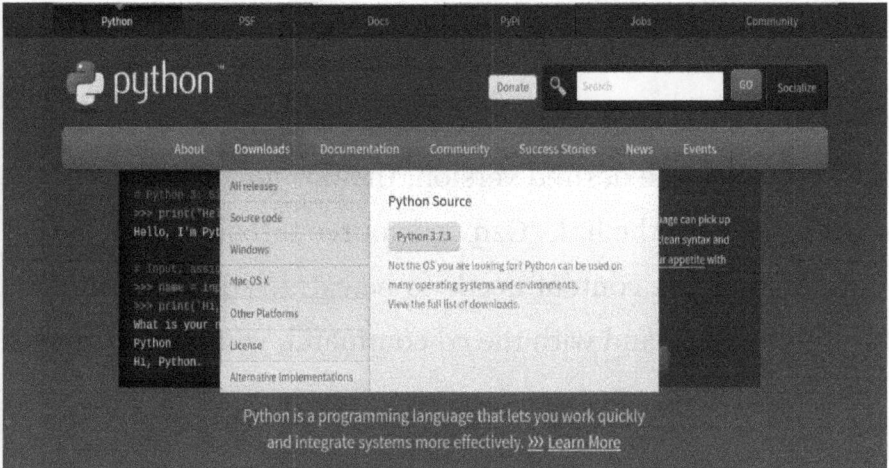

If the version that is needed is the one shown on the right side of the menu, press the button that says, "Python 3.XX", in this case, it is "Python 3.7.3". If you require a different version, press the "Source code" button. In this section, it will load a screen that contains the links to download the different versions of Python that exist.

Once you locate the desired version, the download process of the file will start with the link "Gzipped source tarball." After the file is downloaded, the content needs to be extracted. The command interpreter opens, and with the cd command, you have to change to the extracted directory.

cd / directory / Python-XXX

Once in the directory, you have to execute the installer configuration file.

./configure

Then, you should compile the file with the make command, and test the system compatibility with make test.

make

make test

After this, it is necessary to know how the Python version will be installed. It is possible to define the version to be installed as the main version of the system, and it is possible to install it as an alternative version. It is not recommended to install it as the main version since problems in the system may be caused by compatibility problems between the different versions of Python. In case you want to install as the main version, you will need to execute the following command:

sudo make install

If the version is to be installed as an alternative version (recommended), then you have to execute the following command:

sudo make altinstall

In this way, Python will be installed on the system. It should be noted that depending on the version of Python that is installed (if it is installed as an alternative version), the way to start the Python interpreter will change. If the system has Python 3.7 by default, and Python 3.6 is installed as an alternative, to start the Python 3.6 interpreter, the command needs to be executed is:

python3.6

The way to start Python 3.7 will remain the same.

macOS

macOS is the operating system of the personal computers of Apple Inc. It is widely used throughout the world because of the great efficiency it has. Installing Python on these systems, as in Linux, is generally not necessary, as it comes with pre-installed version 2.7. However, a version different from the one installed by default in the system can be installed. To do this, there are different paths, among which one is the installation process detailed above for Linux. Here are the details in the simplest way that does not require extra software to install Python on macOS:

At the same facilities in Linux and Windows, the process begins by accessing the section downloads of the (macOS) operating system of Python.org and choosing the desired version.

Python Releases for Mac OS X

- Latest Python 3 Release - Python 3.7.3
- Latest Python 2 Release - Python 2.7.16

Stable Releases

- Python 3.8.9 - July 2, 2019
  - No files for this release
- Python 3.7.3 - March 25, 2019
  - Download macOS 64-bit/32-bit installer
  - Download macOS 64-bit installer
- Python 3.4.10 - March 18, 2019
  - No files for this release
- Python 3.5.7 - March 18, 2019
  - No files for this release.
- Python 2.7.16 - March 4, 2019
  - Download macOS 64-bit/32-bit installer
  - Download macOS 64-bit installer
- Python 3.7.2 - Dec. 24, 2018
  - Download macOS 64-bit/32-bit installer

Pre-releases

- Python 3.6.9rc1 - June 18, 2019
  - No files for this release.
- Python 3.7.4rc1 - June 18, 2019
  - Download macOS 64-bit/32-bit installer
  - Download macOS 64-bit installer
- Python 3.8.0b1 - June 4, 2019
  - Download macOS 64-bit installer
- Python 3.8.0a4 - May 6, 2019
  - Download macOS 64-bit installer
- Python 3.8.0a3 - March 25, 2019
  - Download macOS 64-bit installer
- Python 3.7.3rc1 - March 12, 2019
  - Download macOS 64-bit installer

In this case, Python 3.6.0 will be installed with the automatic installer. The process is started by executing the downloaded file. A screen with information about the installation package will start. Proceed by pressing the "Continue" button.

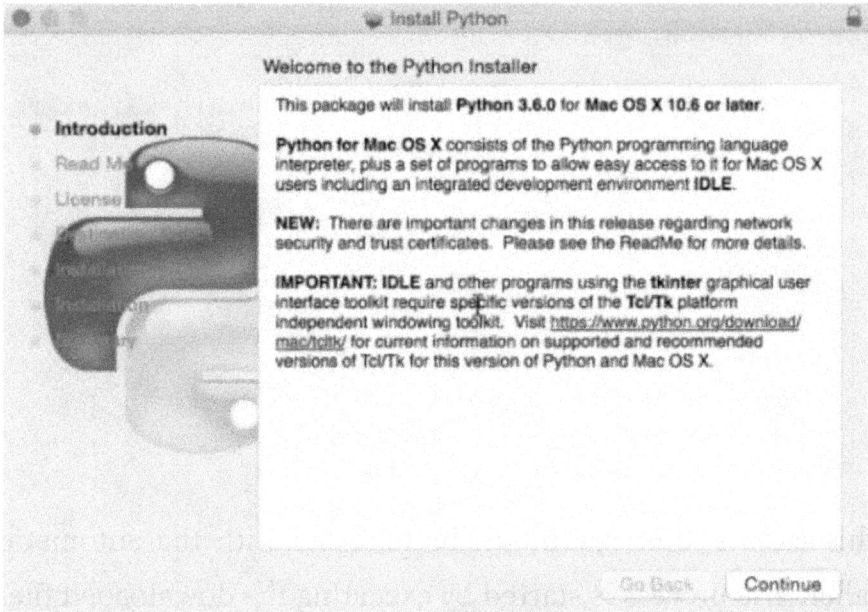

Welcome to the Python Installer

This package will install **Python 3.6.0** for **Mac OS X 10.6 or later**.

Python for Mac OS X consists of the Python programming language interpreter, plus a set of programs to allow easy access to it for Mac OS X users including an integrated development environment **IDLE**.

NEW: There are important changes in this release regarding network security and trust certificates. Please see the ReadMe for more details.

IMPORTANT: IDLE and other programs using the **tkinter** graphical user interface toolkit require specific versions of the **Tcl/Tk** platform independent windowing toolkit. Visit https://www.python.org/download/mac/tcltk/ for current information on supported and recommended versions of Tcl/Tk for this version of Python and Mac OS X.

On the next screen, you will find the "Read Me" and the License information. In both steps, accept the license agreement by pressing "Agree" and then press the "Continue" button.

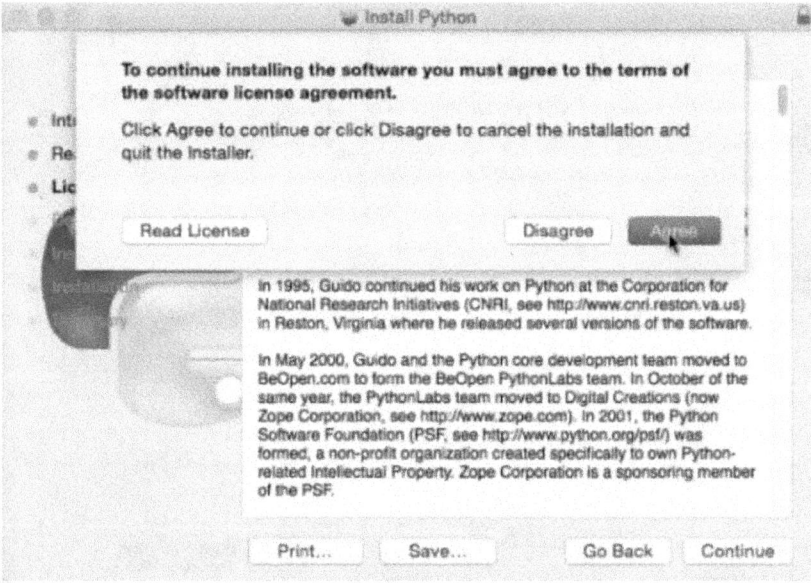

After this, information appears regarding the type of installation that will be done, and the weight of the software here clicks on the "Install" button.

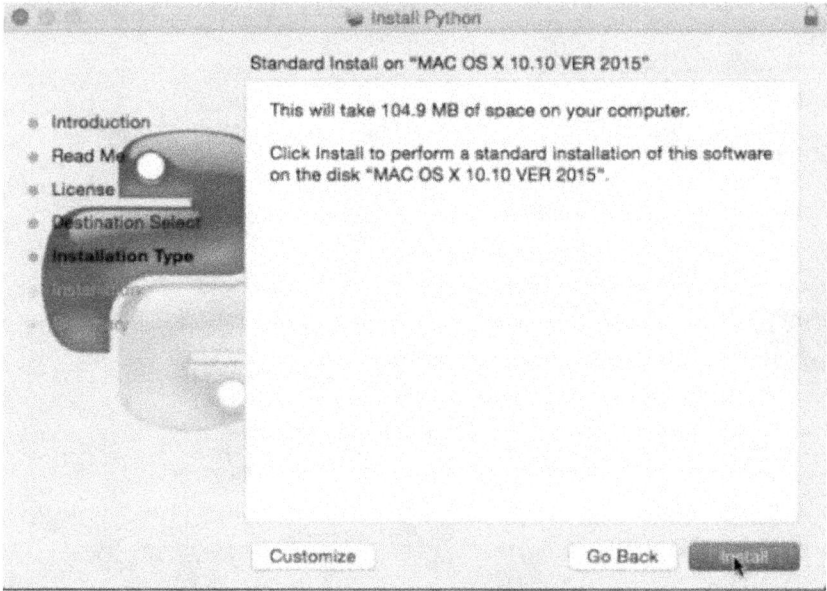

Install Python

Standard Install on "MAC OS X 10.10 VER 2015"

- Introduction
- Read Me
- License
- Destination Select
- Installation Type

This will take 104.9 MB of space on your computer.

Click Install to perform a standard installation of this software on the disk "MAC OS X 10.10 VER 2015".

Customize   Go Back   Install

After entering the user credentials, the installation process will begin.

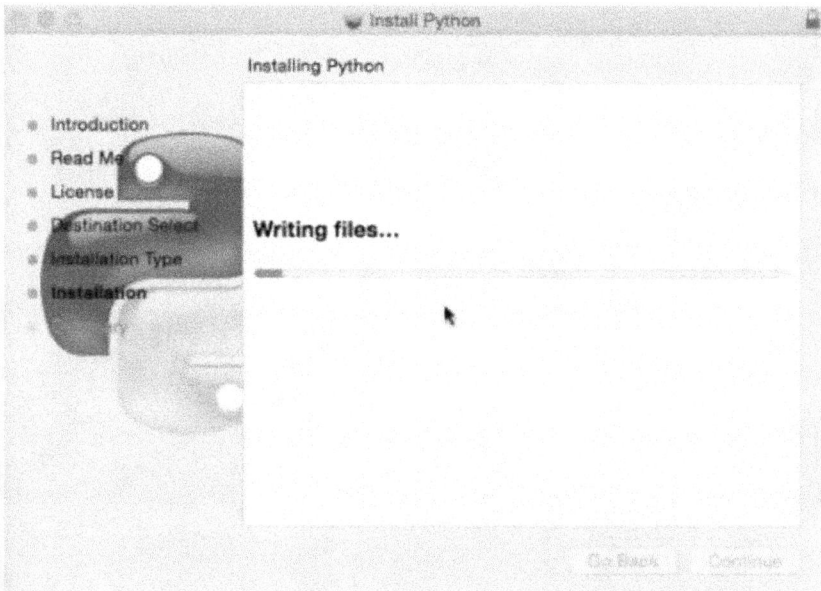

Installing Python

- Introduction
- Read Me
- License
- Destination Select
- Installation Type
- Installation

Writing files...

Go Back   Continue

Upon reaching 100%, the installation will be finished, and Python will already be installed on the system.

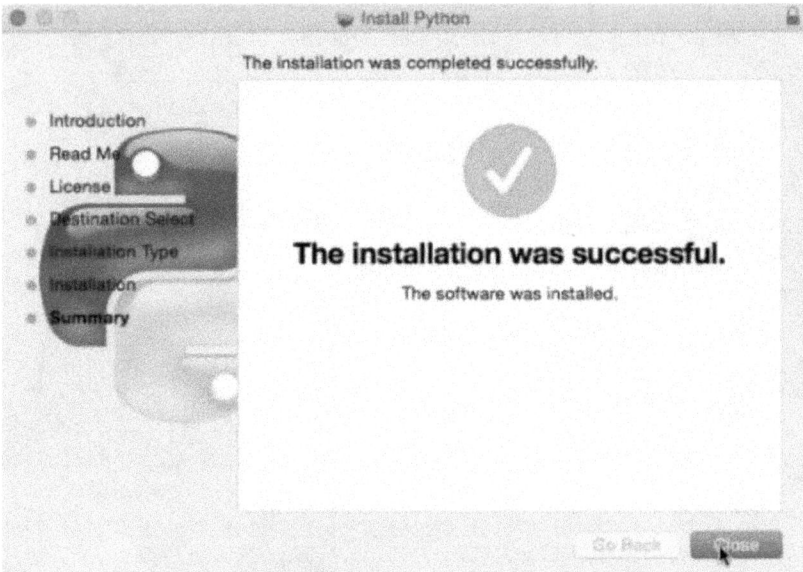

The installation was completed successfully.

The installation was successful.

The software was installed.

After following some of these ways to install Python, depending on your operating system, you will have the right environment to start analyzing your Time-series or other projects in Python.

Now that we have Python installed and working, let's get going. Let's start off with the reading of time-series data using Python and calculation of statistics (graphing and autocorrelation). I will explain the process of automatic SARIMA model estimation. After reading this, you will learn how to implement the simple Box-Jenkins method in Python.

Time-series analysis is, as the name implies, a method of analyzing time-series data. Time-series data is the data that increases daily (or weekly, monthly, yearly), such as "daily sales data," "daily temperature data," and "monthly airplane passenger count." Time-series data often has a relationship, such as "Yesterday's sales are similar to today's sales."

So, if you use time-series data well, you might be able to predict future sales data from yesterday's sales data. By learning time-series analysis, you can create models that predict the future from the past.

Here, we will explain how to use time-series analysis with popular Python and how to predict the future through modeling. This section deals with the Box-Jenkins method, mainly using the ARIMA model.

Reading Time-Series Data

First, we are going to load the required libraries at once, as shown in the image below.

```
# Load basic library
import numpy as np
import pandas as pd
from scipy import stats

# Draw graph
from matplotlib import pylab as plt
import seaborn as sns
% matplotlib inline
sns.set ()

# Make the graph landscape
from matplotlib.pylab import rcParams
rcParams [ 'figure.figsize' ] = 15, 6

# Statistical model
import statsmodels.api as sm
```

This time, we'll cover data on the number of airline passengers per month. The period is from January 1949 to December 1960.

The first appearance is "Box, GEP, Jenkins, GM and Reinsel, GC (1976) Time-series analysis, Forecasting and Control, Third Edition, Holden-Day, Series G." You can check the details of the data here.

Data Link: https://stat.ethz.ch/R-manu;

devel/library/datasets/html/AirPasseng

Also, the data can be found as a CSV file from this site, and here you can download the file.

CSV Link:

https://www.analyticsvidhya.com/blog/2016/02/ti

forecasting-codes-python/

Download Link:

httns·//www analvticsvidhva com/wn-

The first column of this data is the date. Please note that if you open and save it in Excel, it will be converted to a format you would not understand well. To open the file, use a text editor such as Notepad.

Now let's load the data first in the normal reading mode.

```
# Read data normally
# https://stat.ethz.ch/R-manual/R-devel/library/datasets/html/AirPassengers.html
dataNormal = pd.read_csv ('AirPassengers.csv')
dataNormal.head ()
```

You can read the data, but in this way, the first column is treated as "data." The first column is just a date. So, it is necessary to specify and read, "The first column is a date index." In addition, the SARIMA model will be estimated later, but at this time, if the data is int type (integer), an error will occur, as follows.

```python
# Read in date format (If you do not read with dtype = float, an error will occur later when estimating the ARIMA model)
dateparse = lambda dates: pd.datetime.strptime (dates, '%Y-%X m')
data = pd.read_csv ('AirPassengers.csv', index_col = 'Month', date_parser = dateparse, dtype = 'float')
data.head ()
```

If the date format is simple, it is easier to read as follows.

```python
# The following code is easier to read
data = pd.read_csv ('AirPassengers.csv',
                    index_col = 'Month',
                    parse_dates = True,
                    dtype = 'float')
data.head ()
```

As you can see from the JupyterNotebook results, the first column is just a date index. The data is a monthly basis with the number of days added to it (all are assumed to be the first day). This makes it easier to treat as a date.

Since the data is only one column, there is no need to keep it in the data frame. Extract only the series of passenger numbers and store it in a variable named "ts." From now on, we will proceed with the analysis using this variable called "ts."

```
# Use date format
ts = data ['# Passengers']
ts.head ()
```

Handling of Time-Series Data

First, plot the data. If you do not draw a graph, data analysis will not begin. It can be plotted with the following one line.

plt.plot(ts)

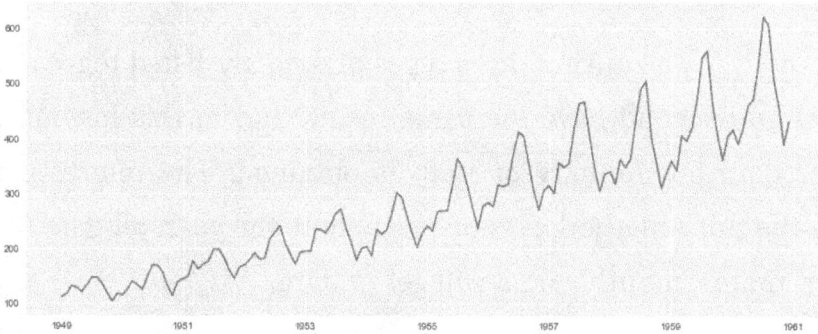

When you draw a graph, you can understand many things by yourself from it. First, the number of passengers is increasing year by year. Then, the number of passengers is likely to change periodically in each season. I will model the presence or absence of seasonal variation later. First, let's get used to handling time-series data.

93

To get data for a specific year and month, you can try the following:

```python
# Data acquisition method # 1
ts ['1949-01-01']

# Data acquisition method # 2
from datetime import datetime
ts [datetime (1949,1, 1)]

# Get all the data of 1949
ts ['1949']
```

The methods for acquiring three types of data are listed together. The first and second have the same result, and in this instance, only the data for January of 1949 is obtained. The interesting thing is the third method, if you specify only the year, all data for the year, that is, monthly data will get 12 data.

The next step is how to shift the data and take the difference.

Literally, shifting means "shifting the data," and by using the shifting, we can quickly obtain the difference between the data. For example, if you want to find out how many passengers increased in February of 1949 compared to January 1949, you can calculate the difference.

In the case of time-series analysis, logarithmic differences are often taken. The logarithmic difference is an index that roughly represents the "rate of change." Another advantage is that taking the logarithm makes the data easier to fit the model. This time, we do not use the logarithmic difference series, but just check the calculation method.

Here is a summary of how to calculate shift, difference, and logarithmic difference.

Shift applies the "shift ()" function: ts.shift ()

If you take out only your head, it looks like this: ts.shift ().head ()

Here is the result:

```
Month
 1949-01-01 NaN
 1949-02-01 112.0
 1949-03-01 118.0
 1949-04-01 132.0
 1949-05-01 129.0
 Name: Passengers, dtype: float64
```

The first data point is NaN because there is no value prior to it. The second data point (112.0) was the first point. The third (118.0) was the second, and so on.

Now, the difference between the first data set and second data set is calculated to get the actual change:

diff = ts − ts.shift ()

diff.head ()

```
Month
1949-01-01 NaN
1949-02-01 6.0
1949-03-01 14.0
1949-04-01 -3.0
1949-05-01 -8.0
Name: Passengers, dtype: float64
```

The logarithmic difference is simply taking the logarithm and then subtracting.

logDiff = np.log (ts) − np.log (ts.shift ())

If you are worried about "NaN" appearing in the result, apply the "dropna" function as follows.

logDiff.dropna ()

```
Month
1949-02-01      0.052186
1949-03-01      0.112117
1949-04-01     -0.022990
1949-05-01     -0.064022
1949-06-01      0.109484
Name: #Passengers, dtype: float64
```

Autocorrelation Coefficient Estimation

"Autocorrelation" shows how similar the previous term and the current term are. If you have a positive autocorrelation, you know that if there were more passengers last month, there would be more this month. The opposite is true for negative autocorrelation.

Autocorrelation in the statistics means that the values / measured values of a variable are correlated with themselves. This is regularly the case if the measured values of a variable are recorded or measured over time.

For example, a newspaper publisher records the number of subscribers every month. In January, for example, there were 100,000 subscribers. In February, there were 102,000. There is a connection between the individual results: a large proportion of the subscribers counted in February were already available in January (some have probably canceled; others have been gained). The number of subscribers in February, therefore, depends on the number in January, and it is the same in the other months (the number of subscribers in March depends on that in February, etc.).

To understand better, assume that the unemployment statistics show 3.56 million unemployed in March, in April it is 3.48 million and in May 3.42 million. These values are related. A large number of the unemployed from March are still without work in April and also in May - there are not suddenly 3, x million other people unemployed. Therefore, the number of unemployed in a month is always related to the number of the previous month - there is autocorrelation.

A counterexample is a random number in roulette. Whether the 36, 17, or 4 fell earlier does not affect the subsequent run of the ball. If you look at the fallen roulette numbers over night as a diagram, you cannot draw a wavy curve like with the unemployment statistics. Instead, the view falls on a jagged curve that alternates wildly between the numbers 0 to 36.

So, let's look at an autocorrelation example here. We are going to look at the airline passenger data from the previous example and do a autocorrelation analysis in the code below:

```
# Find autocorrelation
ts_acf = sm.tsa.stattools.acf (ts, nlags = 40)
ts_acf

# Partial autocorrelation
ts_pacf = sm.tsa.stattools.pacf (ts, nlags = 40, method = 'ols')
ts_pacf
```

The results are long:

```
array([ 1.        ,  0.95893198, -0.32983096,  0.2018249 ,  0.14500798,
        0.25848232, -0.02690283,  0.20433019,  0.15607896,  0.56860841,
        0.29256358,  0.8402143 ,  0.61268285, -0.66597616, -0.38463943,
        0.0787466 , -0.02663483, -0.05805221, -0.04350748,  0.27732556,
       -0.04046447,  0.13739883,  0.3859958 ,  0.24203808, -0.04912986,
       -0.19599778, -0.15443575,  0.04484465,  0.18371541, -0.0906113 ,
       -0.06202938,  0.34827092,  0.09899499, -0.08396793,  0.36328898,
       -0.17956662,  0.15839435,  0.06376775, -0.27503705,  0.2707607 ,
        0.32002003])
```

The positive and negative values we can see from the above image are the coefficients (results of autocorrelation) in different lags. These are basically the correlation coefficients between the current and previous values.

A positive autocorrelation indicates that an increase in time series is often followed by another increase. If the autocorrelation result is close to 1, it is almost certain to be followed by another increase. In other words, the average time series value is increasing. On the other hand, a decrease is almost undoubtedly followed by another decrease. In other words, the average time series value is decreasing. Part of the "trend" follows trivially.

You can also draw a graph of autocorrelation.

```
# Autocorrelation graph
fig = plt.figure (figsize = (12,8))
ax1 = fig.add_subplot (211)
fig = sm.graphics.tsa.plot_acf (ts, lags = 40, ax = ax1)
ax2 = fig.add_subplot (212)
fig = sm.graphics.tsa.plot_pacf (ts, lags = 40, ax = ax2)
```

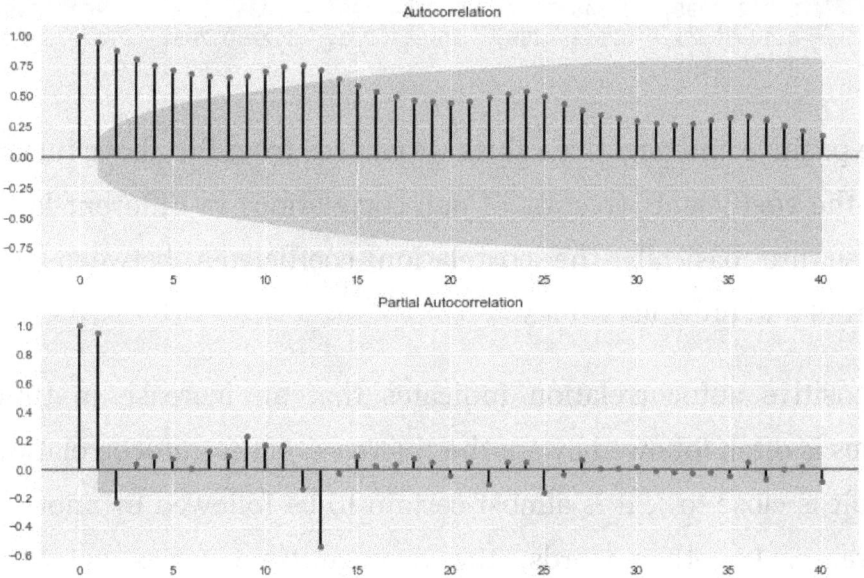

1 or 12 months is the maximum lag, indicating a positive correlation with the 12-month cycle.

There is a clear seasonal pattern and an upward trend. It is not fair enough to claim that the time series is in the stationary pattern. The population ACF has only been defined for a stationary time series model; the ACF sample can be calculated for any time series.

Due to seasonality, autocorrelation at the end of lag 12 is high, and for each multiple of 12, autocorrelation will tend to be high but will continue to decrease.

ARIMA Model Estimation

For Python, some functions automatically determine the ARMA order, but for some reason, it only identifies the "ARMA" order. In other words, the number of times of the difference is not automatically determined.

Well, this data looks like a summation process at a glance, so I will analyze it after taking the difference. The decision on whether to make the difference is discussed again in the last section, "Automatic SARIMA Model Estimation." In summary, you have to do the 'for' loop yourself to find the optimal order.

Here, the difference is determined once with a fixed hit, and then the automatic ARMA order determination function is applied.

```python
# Probably the difference process
diff = ts-ts.shift ()
diff = diff.dropna ()

# Execute automatic ARMA estimation function on difference series
resDiff = sm.tsa.arma_order_select_ic (diff, ic = 'aic', trend = 'nc')
resDiff
```

The creation of the difference series has already been explained earlier, so it is avoided here. Just note that we are removing the annoying NaN with dropna. You can determine the ARMA order by using the function "arma_order_select_ic."

The arguments are the target data, what to use as the information criterion (otherwise BIC can be used), and the presence or absence of a trend. The original series seemed to have a pattern, but when I took the difference and looked, it seemed that there was no trend, so I did not include it this time.

Now that we know the optimal order, we can rebuild the model.

```
# Model P-3, q = 2 because it is best
from statsmodels.tsa.arima_model import ARIMA
ARIMA_3_1_2 = ARIMA (ts, order = (3, 1, 2)). Fit (dist = False)
ARIMA_3_1_2.paramsm.tsa.arma_order_select_ic (diff, ic = 'aic', trend = 'nc')
resDiff
```

```
const  2.673501
ar.L1.D.#Passengers  0.261992
ar.L2.D.#Passengers  0.367827
ar.L3.D.#Passengers  -0.363472
ma.L1.D.#Passengers  -0.075005
ma.L2.D.#Passengers  -0.924805
dtype: float64
```

However, this model has its drawbacks, as this has not been able to reflect periodic seasonal variations so well, but the autocorrelation of the residuals is clear at a glance.

```
# Check for residuals
# Because it is not SARIMA, the periodicity remains. . . .
resid = ARIMA_3_1_2.resid
fig = plt.figure (figsize = (12,8))
ax1 = fig.add_subplot (211)
fig = sm.graphics.tsa.plot_acf (resid.values.squeeze (), lags = 40, ax = ax1)
ax2 = fig.add_subplot (212)
fig = sm.graphics.tsa.plot_pacf (resid, lags = 40, ax = ax2)
```

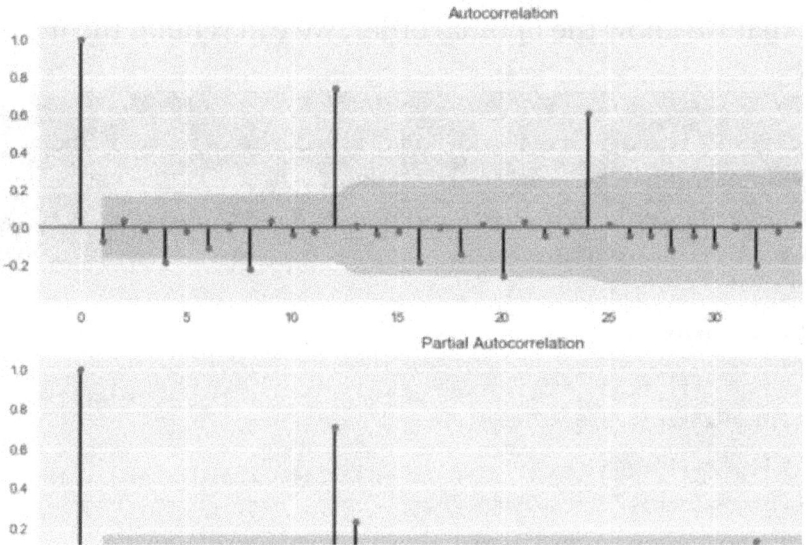

Remember that the PACF can be used to find the best AR model sequence. The gray shaded horizontal area represents important factors. The vertical spikes express the ACF and PACF values at the relevant time. Only vertical lines that cross the edge of the horizontal gray shade are considered significant.

SARIMA Model Estimation

Before estimating the SARIMA model, there is a caveat. The SARIMA model is calculated using a library called "statsmodels," which we will discuss in the next part of this chapter. However, the SARIMA model is not included unless the version is 0.8.0 or higher.

If you get angry with Python when trying to estimate a SARIMA model, you can upgrade your version of statmodels. If you are using Anaconda on Windows, launch the command prompt, and execute the following command.

conda install -c taugspurger statsmodels = 0.8.0

Now we are ready to estimate the SARIMA model. For seasonal variation, the order is fixed.

```
# Estimate the SARIMA model by "fixing"
import statsmodels.api as sm

SARIMA_3_1_2_111 = sm.tsa.SARIMAX (ts, order = (3,1,2), seasonal_order = (1,1,1,12)). Fit ()
print (SARIMA_3_1_2_111.summary ())
```

Use the function "SARIMAX." The name has an "X" in it, which means that other external variables can be included in the model, as in regression analysis.

This time we will go with only one variable. Set the order of the ARIMA model with order = (3, 1, 2) and the order of seasonal variation with seasonal_order = (1,1,1,12).

The last "12" in seasonal_order = (1, 1, 1, 12) means "12-month cycle". So, in effect, (sp, sd, sq) = (1, 1, 1).

The result is shown below. During the calculation, I got a warning that the result of the maximum likelihood method did not converge but ignore it for the time being and proceed.

```
                           Statespace Model Results
==============================================================================
Dep. Variable:                    Passengers   No. Observations:          144
Model:           SARIMAX(3, 1, 2)x(1, 1, 1, 12)  Log Likelihood        -502.990
Date:                     Sat, 27 May 2017   AIC                     1021.980
Time:                             18:50:26   BIC                     1045.738
Sample:                         01-01-1949   HQIC                    1031.634
                              - 12-01-1960
Covariance Type:                       opg
==============================================================================
                 coef    std err          z      P>|z|      [0.025      0.975]
------------------------------------------------------------------------------
ar.L1          0.5383      1.665      0.323      0.747      -2.725       3.802
ar.L2          0.2830      0.969      0.292      0.770      -1.617       2.183
ar.L3         -0.0346      0.407     -0.085      0.932      -0.832       0.762
ma.L1         -0.9231      1.690     -0.546      0.585      -4.236       2.390
ma.L2         -0.0519      1.658     -0.031      0.975      -3.302       3.198
ar.S.L12      -0.8773      0.285     -3.077      0.002      -1.436      -0.318
ma.S.L12       0.7839      0.370      2.116      0.034       0.058       1.510
sigma2       124.6735     14.315      8.709      0.000      96.616     152.731
==============================================================================
Ljung-Box (Q):                       51.03   Jarque-Bera (JB):          13.51
Prob(Q):                              0.11   Prob(JB):                   0.00
Heteroskedasticity (H):               2.60   Skew:                       0.14
Prob(H) (two-sided):                  0.00   Kurtosis:                   4.55
==============================================================================
```

The first thing that stands out is the header called "Statespace Model (SSM) Results," and it seems that you and the time before the statespace model came out, but it seems that you are using a Kalman filter internally for parameter estimation.

By using the statespace model, the Box-Jenkins method can be handled uniformly. You probably use this for actual calculations. The summary results show a list of information criteria, such as AIC, as well as estimated coefficients. If it is said to be a perfect result, unfortunately, it does not, but we will proceed as it is.

The autocorrelation of the residuals should be almost delicate.

```
# Check for residuals
residSARIMA = SARIMA_3_1_2_111.resid
fig = plt.figure (figsize = (12,8))
ax1 = fig.add_subplot (211)
fig = sm.graphics.tsa.plot_acf (residSARIMA.values.squeeze (), lags = 40, ax = ax1)
ax2 = fig.add_subplot (212)
fig = sm.graphics.tsa.plot_pacf (residSARIMA, lags = 40, ax = ax2)
```

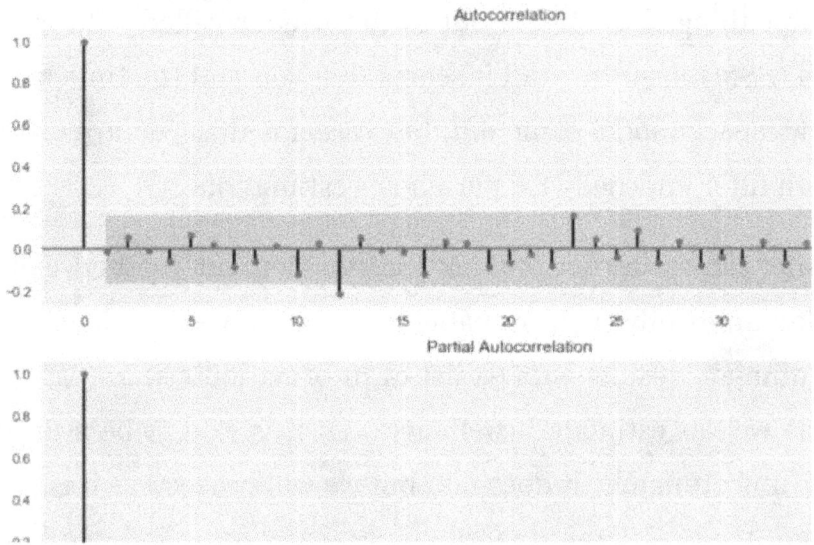
Autocorrelation

Partial Autocorrelation

Now we have a model, so let's make a prediction.

```
# Forecast
pred = SARIMA_3_1_2_111.predict ('1960-01-01', '1961-12-01')

# Illustrate actual data and forecast results
plt.plot (ts)
plt.plot (pred, "r")
```

We use the "predict (forecast start date, forecast end date)"
function for the forecast as it is, but note that the forecast target
date must start from the "date included in the original sample."

Data is available through December 1960.

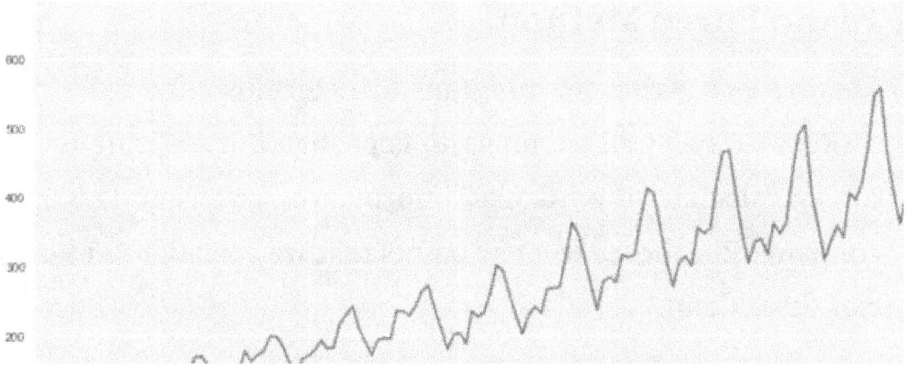

Dynamics = the false argument causes the predictions to be made in one step, i.e., the predictions are made at each point using the complete history up to that point.

However, it is possible to obtain a reasonable representation of our predictive power by using dynamic predictions. In this case, we only use time series information up to a certain point, and the predictions are then made using values from previous prediction data. Let's try the dynamic prediction calculation.

Determination of SARIMA Model Order by Brute Force Method

Finally, let's write our program to determine the order of the SARIMA model; all we must do is change the ordering by brute force, calculate the Akaike information criterion (AIC), and compare. But there are a few points that we can add, so I hope we can share them.

The Akaike information criterion is a means of the goodness of fit of a statistical model. It can indeed be pointed out that it defines the relationship between bias and variance within the structure of the model, or with regards to the accuracy and complexity of the model.

The AIC is not a model analysis in the perception of hypothesis testing. Instead, it generates a method for the divergence between the models of a tool for model options. Provided a set of facts, different options of models could be categorized depending on their AIC by using the model that has the minimal AIC, which is the best.

First, prepare a container for storing the AIC. If you decide the maximum order, you can calculate and calculate how many models of the model will be created. Therefore, calculate the number of repetitions in advance and secure the number of lines by that amount.

The following rules describe the degree:

ARIMA (p, d, q)

Season (sp, sd, sq)

Where:

- p is the degree of the autoregressive model: AR (p),
- q is the degree of the moving average model: MA (q), and
- d is the number of times to make the difference: I (d).

```
# Find SARIMA order that minimizes AIC by brute force
max_p = 3
max_q = 3
max_d = 1
max_sp = 1
max_sq = 1
max_sd = 1

pattern = max_p * (max_q + 1) * (max_d + 1) * (max_sp + 1) * (max_sq + 1) * (max_sd + 1)

modelSelection = pd.DataFrame (index = range (pattern), columns = ["model", "aic"])
pattern
```

When you do this, the number of patterns will be "192". That is, the SARIMA model is estimated 192 times repeatedly. Setting the maximum order to "3" is just a rule of thumb. Just keep in mind that increasing this takes more time.

Regarding the degree of the season, we have a small image because there is an image that does not increase. If you are confident in your computer resources, set a higher number. In the formula for calculating the number of patterns (line 9), max_p is not incremented by one, but this is not a typo.

When the AR term of the ARIMA model was executed with the order set to 0, an error occurred frequently. Therefore, the AR term was looped on the condition that the minimum order was set to 1, so the calculation can be performed only for such slightly distorted times.

Here is the code to actually estimate the SARIMA model, just turn the loop.

```
# Automatic SARIMA selection
num = 0

for p in range (1, max_p + 1):
    for d in range (0, max_d + 1):
        for q in range (0, max_q + 1):
            for sp in range (0, max_sp + 1):
                for sd in range (0, max_sd + 1):
                    for sq in range (0, max_sq + 1):
                        sarima = sm.tsa.SARIMAX (
                            ts, order = (p, d, q),
                            seasonal_order = (sp, sd, sq, 12),
                            enforce_stationarity = False,
                            enforce_invertibility = False
                        ) .fit ()
                        modelSelection.ix [num] ["model"] = "order = (" + str (p) + "," + str (d) + "," + str (q) + "), season = (" + str
(sp) + "," + str (sd) + "," + str (sq) + ")"
```

Python does not use curly braces and expresses loops with indentation, but if you nest so many loops so far, it is a bit hard to read. Review the arguments used in the SARIMAX function. Lines 13 and 14 specify "enforce_stationarity = False" and "enforce_invertibility = False", respectively.

Note that without this, you will get lots of errors and would not be able to turn the loop.

When applying a normal autoregressive model to the summation process, "the model of the stationary process must be estimated as → enforce_stationarity = True" and specify that the model cannot be estimated in the first place. Usually, when dealing with the summation process, the autoregressive model is always applied after taking the difference; therefore, to eliminate the error, this specification can be forcibly specified.

Normally, the result of `` applying the autoregressive model to the integration process " is not selected, but just because it is the smallest AIC does not adopt it as a release, but whether it has a `` decent " order, so check it before using. "Enforce_invertibility" specifies whether to maintain the invertible condition of the moving average model. If this is not set to False, there will be many errors.

In addition, although the error no longer appears, a large number of warnings appear that the parameter estimation result is "not converged." The 15th line of the code, when fitting the model, changing the optimization method as follows + increasing the number of repetitions, reduced the number of warnings. Still, not all, some get warnings.

fit (method = 'bfgs', maxiter = 300, disp = False)

I think that the selected result should not be over trusted but must be analyzed with some doubt. Results come out in minutes. If you pick up the smallest AIC model, it will look like this.

```
# AIC smallest model
modelSelection [modelSelection.aic == min (modelSelection.aic)]
```

model aic

187 order = (3,1,3), season = (0,1,1) 898.105

order = (3,1,3), season = (0,1,1) became the AIC minimum model, so rebuild the SARIMA model again using this order.

```
#
#
bestSARIMA = sm.tsa.SARIMAX(ts, order=(3,1,3), seasonal_order=(0,1,1,12), enforce_stationarity = False, enforce_invertibility = False).fit()
print(bestSARIMA.summary())
```

Here is the result.

```
                          Statespace Model Results
==================================================================================
Dep. Variable:                    Passengers   No. Observations:             144
Model:           SARIMAX(3, 1, 3)x(0, 1, 1, 12)   Log Likelihood          -441.052
Date:                       Sat, 27 May 2017   AIC                        898.105
Time:                               18:53:24   BIC                        921.863
Sample:                           01-01-1949   HQIC                       907.759
                                - 12-01-1960
Covariance Type:                         opg
==================================================================================
                 coef    std err          z      P>|z|      [0.025      0.975]
----------------------------------------------------------------------------------
ar.L1         -0.2231      0.097     -2.302      0.021      -0.413      -0.033
ar.L2         -0.1642      0.108     -1.515      0.130      -0.377       0.048
ar.L3          0.7244      0.094      7.704      0.000       0.540       0.909
ma.L1         -0.0837    121.897     -0.001      0.999    -238.997     238.829
ma.L2          0.1221    152.640      0.001      0.999    -299.046     299.290
ma.L3         -0.9797    266.561     -0.004      0.997    -523.429     521.469
ma.S.L12      -0.1583      0.118     -1.337      0.181      -0.390       0.074
sigma2       119.6719   3.26e+04      0.004      0.997    -6.37e+04    6.39e+04
==================================================================================
Ljung-Box (Q):                     36.68   Jarque-Bera (JB):             4.39
Prob(Q):                            0.62   Prob(JB):                     0.11
Heteroskedasticity (H):             1.87   Skew:                         0.16
Prob(H) (two-sided):                0.06   Kurtosis:                     3.90
==================================================================================
```

The residuals do not seem to be a big problem.

```
# Check for residuals
residSARIMA = bestSARIMA.resid
fig = plt.figure (figsize = (12,8))
ax1 = fig.add_subplot (211)
fig = sm.graphics.tsa.plot_acf (residSARIMA, lags = 40, ax = ax1)
ax2 = fig.add_subplot (212)
fig = sm.graphics.tsa.plot_pacf (residSARIMA, lags = 40, ax = ax2)
```

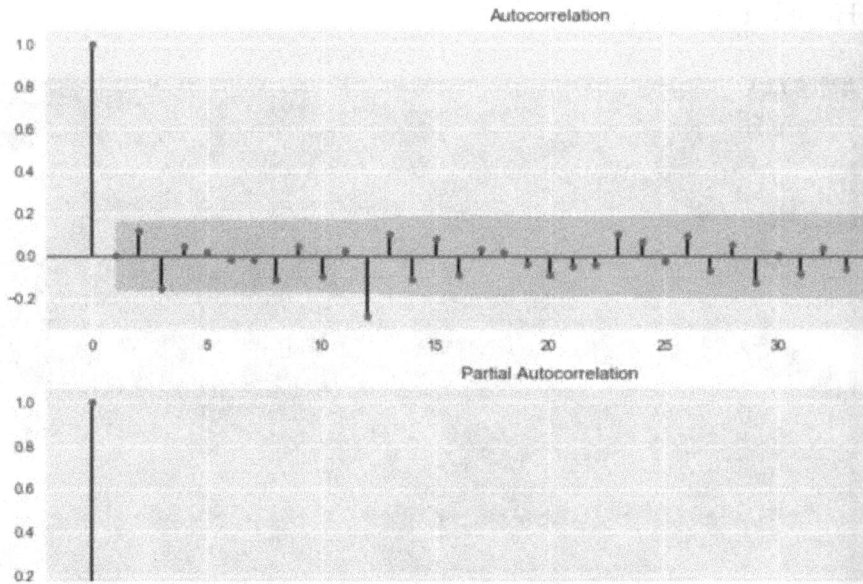

Autocorrelation

Partial Autocorrelation

The last is the prediction.

```
# Forecast
bestPred = bestSARIMA.predict ('1960-01-01', '1961-12-01')
# Illustrate actual data and forecast results
plt.plot (ts)
plt.plot (bestPred, "r")
```

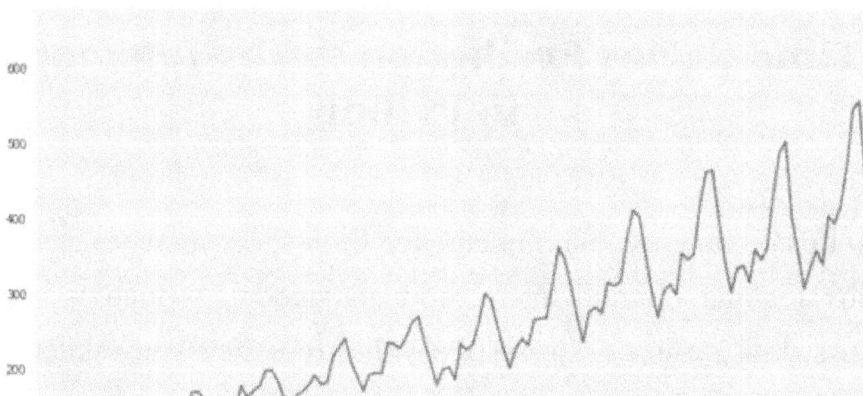

It is important to use a single split when making our training and testing tools because the data MUST guarantee continuity. In fact, the SARIMA model relies on the sequence of the data to make predictions. Looking at the graph above, the model that is ordered as the brute force method does a great job of predicting time series.

Time-Series Prediction With Statsmodels in Python

In this section, we will implement a time-series analysis using the Box-Jenkins method again, and statsmodels, which is also a library of Python. This section will strengthen your skills after getting started with time-series analysis in Python.

A detailed explanation of the method, such as the stationary process and the type of unit root test, will not be explained here because we have already covered that. So, in turn, we will continue just with the implementation.

What is Statsmodels?

It is a library that is a convenient method of implementing a statistical model. It looks like scikit-learn, though scikit-learn is better for machine learning, and the statsmodel is better for lower end statistics.

For this part, I wanted data that had a friendly and easy-to-understand trend and seasonality, so when I Googled sales of ice cream in Kanazawa, Japan (the capital of ice cream of the east), it grabbed my attention.

Library used

```python
import statsmodels.api as sm
import pandas as pd
import numpy as np
import requests
import io
from matplotlib import pylab as plt
% matplotlib inline
# Make the graph landscape
from matplotlib.pylab import rcParams
rcParams ['figure.figsize'] = 15, 6
```

The graph should be fixed in a landscape orientation in advance.

Data

```python
URL = "https://drive.google.com/uc?id=1MZMKbSQXeVnlAWijTC_hwFCCxPbYNX-Y"

r = requests.get (URL)
row_data = pd.read_csv (io.BytesIO (r.content))
```

The shared link of google drive with the data used is here.

Link:

https://drive.google.com/uc?id=1MZMKbSQXeVnlAWijTC_hwF
CCxPbYNX-Y

```
row_data.head ()
```

Then we get the output as shown below:

	date	earnings	temperature
0	2014/1/1	396.0	3.9
1	2014/2/1	309.0	3.5
2	2014/3/1	447.0	7.4
3	2014/4/1	520.0	12.4
4	2014/5/1	803.0	18.0

Earnings are the sales of ice cream, and the temperature is the average temperature for the month. We will transform this data frame into a form that can be analyzed.

```
# An error occurs when estimating the model unless it is a float type
row_data.earnings = row_data.earnings.astype ('float64')
row_data.temperature = row_data.temperature.astype ('float64')
# make datetime type and index
row_data.date = pd.to_datetime (row_data.date)
data = row_data.set_index ("date")
```

Check the deformed data frame.

```
data.head ()
```

date	earnings	temperature
2014-01-01	396.0	3.9
2014-02-01	309.0	3.5
2014-03-01	447.0	7.4
2014-04-01	520.0	12.4
2014-05-01	803.0	18.0

Now let's see the whole picture.

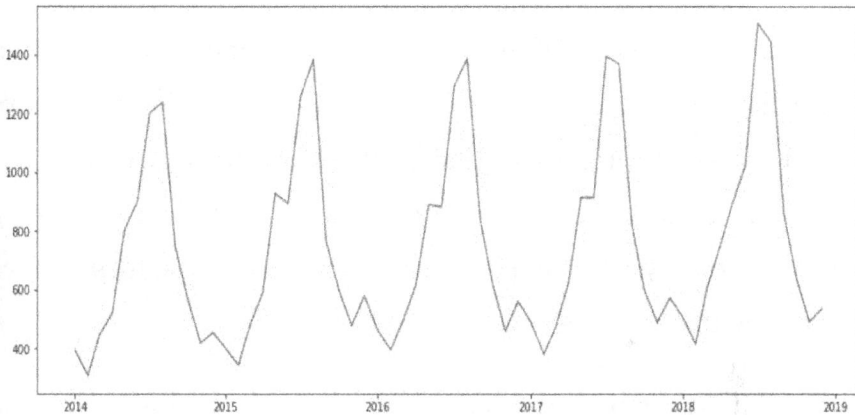

We can see that the data is so easy to understand that its characteristics are clearly visible. Now that the data is ready to process, we will create a model to forecast ice sales. Before that, we will briefly explain some of the important ideas that are necessary for dealing with the time-series data.

Time-series data can be explained by decomposing them into five categories:

1. autocorrelation,
2. periodicity,
3. trend,
4. extrinsic, and
5. white noise.

1. Autocorrelation - The Correlation

The similarity between different points on a time-series is called autocorrelation. Time-series analysis differs significantly from simple regression analysis where it takes into account the relationship factors before and after the data. So, by using autocorrelation, we can express the relationship factors before and after the data is used.

Autocorrelation can be represented by a graph called a correlogram. In statsmodels, you can draw correlograms with graphics plot_acf. But this time, I will not show it because it is similar to autocorrelation, and there is something called "partial autocorrelation." Partial autocorrelation can be plotted with plot_pacf. The autocorrelation of the data this time is shown in the graph below.

```
# Autocorrelation graph
fig = plt.figure (figsize = (12,8))
ax1 = fig.add_subplot (211)
fig = sm.graphics.tsa.plot_acf (data.earnings, lags = 40, ax = ax1)
ax2 = fig.add_subplot (212)
fig = sm.graphics.tsa.plot_pacf (data.earnings, lags = 40, ax = ax2)
```

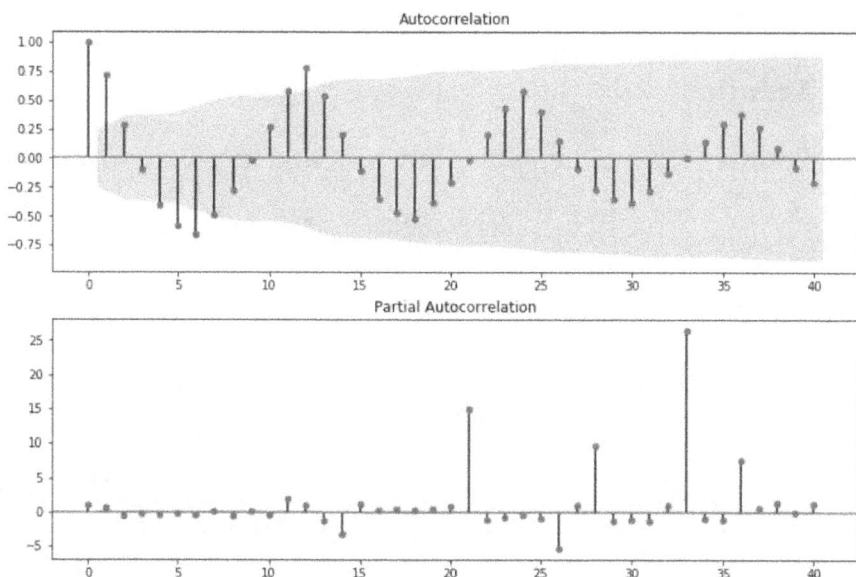

"AutoCorrelation" is the correlogram. By the way, the light blue part indicates the 95% confidence interval. The fact that the data extends outside this confidence interval is statistically significant. The horizontal axis is larger at 12, 24, and 36, indicating that it is similar to the data of 12 months ago. In other words, there is a periodicity every 12 months.

2. Periodicity

Periodicity is also called seasonality, and this refers to a similar change in a 12-month cycle or a periodic change depending on the day of the week. According to the ice cream sales data this time, we can see that there is a seasonality of 12 months cycle.

3. Trends

Trends remain the same, but it is a general trend, in this case, it seems that there is some sort of upward trend.

- ### *Extrinsic*

1 to 3 were the features in the data, but extrinsic refers to data movement that cannot be explained by autocorrelation alone. For example, it seems that the special feature of ice cream on a popular program has made it possible to sell ice cream.

- ### *White noise*

The noise that cannot be explained except for the characteristics of the data described above is called white noise, which is also called as residuals.

Using seasonal_decompose of statsmodels, you can quickly decompose time-series data into trend component, periodic component, and residuals. And you can plot as it is!

```
# Break down data into trends and seasonal components
seasonal_decompose_res = sm.tsa.seasonal_decompose (data.earnings, freq = 12)
seasonal_decompose_res.plot ()
```

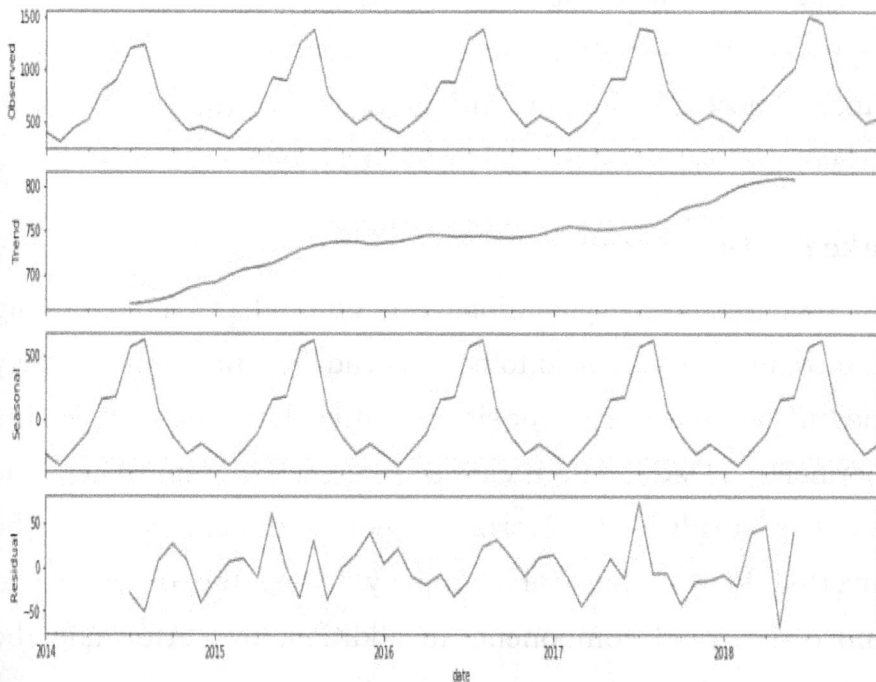

Observed is the original data, and the lower three are the decomposed components. As expected, there seems to be a rising trend and a 12-month periodicity.

Trend and seasonal data can be represented by a model called the "SARIMA" model. Also, this time, the monthly average temperature is prepared as an extrinsic factor, so the "SARIMAX" model that can take into account the extrinsic factor will be adopted. By the way, since the average temperature itself should change almost the same as the periodicity of the sales data, it does not make much sense to include it as honest extrinsic.

Make the Data Steady

Although detailed explanations are omitted, data containing trend components are said to be "unsteady" (data that have many zones in one timestep), making it difficult to analyze. On the other hand, "steady" data is easy to analyze, so I want to keep the data in a steady state. Normally, non-stationary data can be converted to the stationary state by taking the difference to remove the trend component. In addition, by performing the "unit root test," we can check whether the data is steady or not.

Below, we are going to perform unit root tests to analyze the properties of stationarity, keeping in mind that when a series or process has a unit root, the series is not stationary, and the least-squares estimators do not have a normal distribution.

A unit root is a stochastic trend in the time series. Sometimes it is called "random drift ride." Therefore, if the series has a unitary root, it presents a systematic pattern that is unpredictable.

So, a time series is stationary if a change in time does not change the shape of the distribution; and unit roots are a cause of non-stationarity.

As the test, let's do a unit root test on the original series without taking the difference. This time, we used the ADF unit root test. The Augmented Dickey-Fuller (ADF) test is a unit root test for a sample of a time series. It is an extended version of the Dickey-Fuller test for a larger and more complex set of time series models.

The augmented Dickey-Fuller (ADF) used in the analysis is a negative number. The more negative the number is, the higher the rejection of the null hypothesis that there is a unit root for a given level of confidence.

```
# With trend term (up to first order), with constant term
ct = sm.tsa.stattools.adfuller (data.earnings, regression = "ct")
# No trend term, constant term
c = sm.tsa.stattools.adfuller (data.earnings, regression = "c")
# No trend term, no constant term
nc = sm.tsa.stattools.adfuller (data.earnings, regression = "nc")

print ("ct:")
print (ct [1])
print ("-----------------------------------------------------------")
print ("c:")
print (c [1])
print ("-----------------------------------------------------------")
print ("nc:")
print (nc [1])
print ("-----------------------------------------------------------")
```

```
#output
ct:
0.881744298708034
-----------------------------------------------------------
c:
0.8192957538918915
-----------------------------------------------------------
nc:
0.9999999896927978
-----------------------------------------------------------
```

The output value indicates the P-value. In the ADF test, it is a null hypothesis that the data is a unit root process, so if it can be rejected, it can be determined that the data is not a unit root process, which means it is stationary.

The original series is non-stationary because the P-value cannot be rejected in all cases.

Next, shift the data one by one and take the difference, diff () can take the difference of 1 and diff (2) can take the difference of 2 to remove the missing values because they are shifted.

```
diff = data.earnings.diff ()
diff = diff.dropna ()

# With trend term (up to first order), with constant term
ct = sm.tsa.stattools.adfuller (diff, regression = "ct")
# No trend term, constant term
c = sm.tsa.stattools.adfuller (diff, regression = "c")
# No trend term, no constant term
nc = sm.tsa.stattools.adfuller (diff, regression = "nc")

print ("ct:")
print (ct [1])
print ("--------------------------------------------")
print ("c:")
print (c [1])
print ("--------------------------------------------")
print ("nc:")
print (nc [1])
print ("--------------------------------------------")
```

```
# Output
ct:
0.005220302847135729
--------------------------------------------
c:
0.0006567788221081102
--------------------------------------------
nc:
0.0019344470286669219
--------------------------------------------
```

All rejected...!

By the way, if you plot the difference series, it looks like this.

```
plt.plot (diff)
```

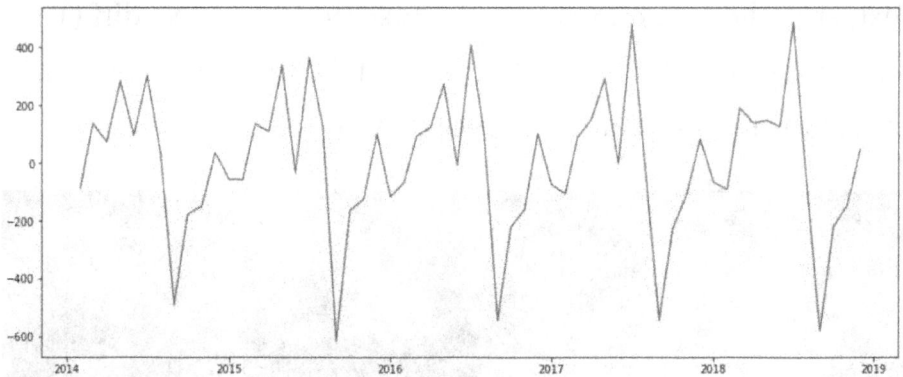

The original series had an upward trend, but the difference series seemed to have a constant average!

Model Estimation

Once the data is steady, all you have to do is estimate the model using SARIMAX in statsmodels. One of the difficulties of the SARIMAX model is that it requires a lot of parameters to set because it can express trends and periodicities. This time, the parameter is estimated semi-automatically using AIC (this is a very common technique).

First, split the data into training and test data.

```
train_data = data [data.index <"2018-06"]
# Test data must include date before test period
test_data = data [data.index> = "2018-05"]
```

Next, estimate the parameters, since the periodicity can be expected to be 12, the other parameters are calculated brute force, and the one with the smallest AIC is used.

```python
# Find SARIMA order that minimizes AIC by brute force
max_p = 3
max_q = 3
max_d = 2
max_sp = 1
max_sq = 1
max_sd = 1

pattern = max_p * (max_d + 1) * (max_q + 1) * (max_sp + 1) * (max_sq + 1) * (max_sd + 1)

modelSelection = pd.DataFrame (index = range (pattern), columns = ["model", "aic"])

# Automatic SARIMA selection
num = 0

for p in range (1, max_p + 1):
    for d in range (0, max_d + 1):
        for q in range (0, max_q + 1):
            for sp in range (0, max_sp + 1):
                for sd in range (0, max_sd + 1):
                    for sq in range (0, max_sq + 1):
                        sarima = sm.tsa.SARIMAX (
                            train_data.earnings, order = (p, d, q),
                            seasonal_order = (sp, sd, sq, 12),
                            enforce_stationarity = False,
                            enforce_invertibility = False
                        ) .fit ()
                        modelSelection.ix [num] ["model"] = "order = (" + str (p) + "," + str (d) + "," + str (q) + "), season = (" + str (sp)
+ "," + str (sd) + "," + str (sq) + ")"
                        modelSelection.ix [num] ["aic"] = sarima.aic
                        num = num + 1
```

This can take several minutes.

After the execution, search the data frame for the one with the smaller AIC.

```python
modelSelection.sort_values (by = 'aic'). head ()
```

	model	aic
278	order=(3,2,2), season=(1,1,0)	265.434
270	order=(3,2,1), season=(1,1,0)	266.521
286	order=(3,2,3), season=(1,1,0)	268.325
254	order=(3,1,3), season=(1,1,0)	269.788
238	order=(3,1,1), season=(1,1,0)	269.873

Since the model with parameters 3, 2, 2, 1, 1, 0 has the smallest AIC, let's use this to estimate the model.

The estimation of the model can be done with only one line below.

```
SARIMA_3_2_2_110 = sm.tsa.SARIMAX (train_data.earnings, order = (3,2,2), seasonal_order = (1,1,0,12)). Fit ()
```

The model is evaluated using the Ljungbox test, and this test is a null hypothesis, which means that "there is no significant autocorrelation in the residuals" (strictly speaking, it is different...), so I am happy if it is not rejected. Significant autocorrelation of the residuals means that there are features of the data that have not yet been accounted for. You can get the residual with model.resid.

134

```
# Ljungbox test
ljungbox_result = sm.stats.diagnostic.acorr_ljungbox (SARIMA_3_2_2_110.resid, lags = 10)
df_ljungbox_result = pd.DataFrame ({"p-value": ljungbox_result [1]})
df_ljungbox_result [df_ljungbox_result ["p-value"] <0.05]
```

I want the output of this code to be an empty data frame because
I do not want to reject it. If AIC is low, it is not necessarily a good
model and not going to help in the analysis.

	p-value
0	0.001934
1	0.005046
2	0.005917
3	0.010949
4	0.018445
5	0.031618
6	0.018867
7	0.000824
8	0.000249
9	0.000466

```
#Model estimation
SARIMA_3_1_3_110 = sm.tsa.SARIMAX (train_data.earnings, order = (3,1,3), seasonal_order = (1,1,0,12)). Fit ()
# Test for residuals
ljungbox_result = sm.stats.diagnostic.acorr_ljungbox (SARIMA_3_1_3_110.resid, lags = 10)
# Result
df_ljungbox_result = pd.DataFrame ({"p-value": ljungbox_result [1]})
df_ljungbox_result [df_ljungbox_result ["p-value"] <0.05]
```

AIC successfully cleared the Ljungbox test with the fourth lowest
parameter!

Try to Predict the Sales

Let's use the resulting model to predict sales during the test period.

```
# Forecast
pred = SARIMA_3_1_3_110.predict ('2018-05-01', '2018-12-01')
# Illustrate actual data and predicted results
plt.plot (train_data.earnings, label = "train")
plt.plot (pred, "r", label = "pred")
plt.legend ()
```

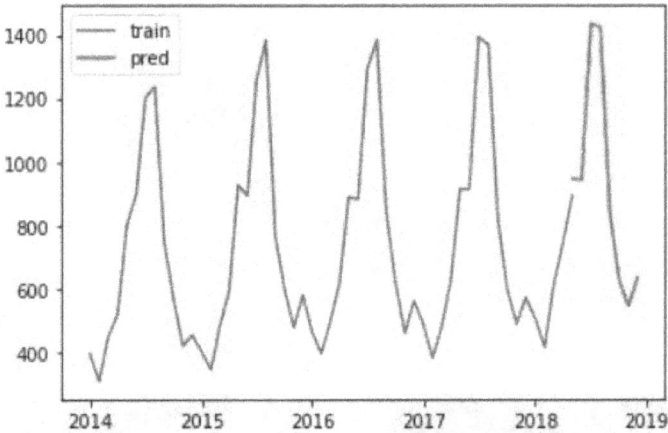

Let's plot together with the correct answer data.

```
# Predictive evaluation
pred = SARIMA_3_1_3_110.predict ('2018-05-01', '2018-12-01')
# Illustrate actual data and predicted results
plt.plot (test_data.earnings, label = "test")
plt.plot (pred, "r", label = "pred")
plt.legend ()
```

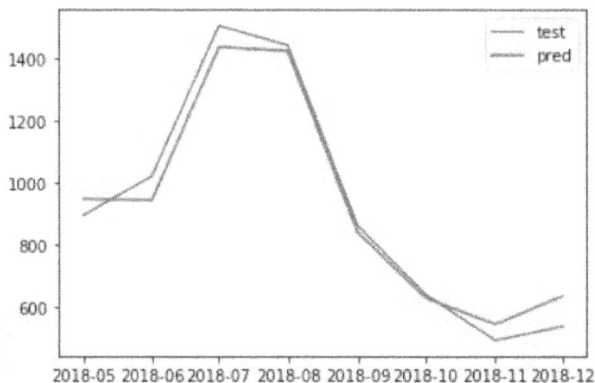

As mentioned earlier, the SARIMAX model can take extrinsic factors into account, in this case, only the average temperature, but it may take multiple extrinsic parameters in different analyses.

```
SARIMA_3_1_3_110_with_exog = sm.tsa.SARIMAX (train_data.earnings, train_data.temperature, order = (3,1,3), seasonal_order = (1,1,0,12)). Fit ()
```

By the way, the result is:

```
# Forecast
# Transform data to (7,1)
test_temperature = [[test_data.temperature [i + 1]] for i in range (7)]
pred_with_exog = SARIMA_3_1_3_110_with_exog.predict ('2018-05-01', '2018-12-01', exog = test_temperature)
# Illustrate actual data and predicted results
plt.plot (test_data.earnings, label = "test")
plt.plot (pred_with_exog, "r", label = "pred_with_exog")
plt.legend ()
```

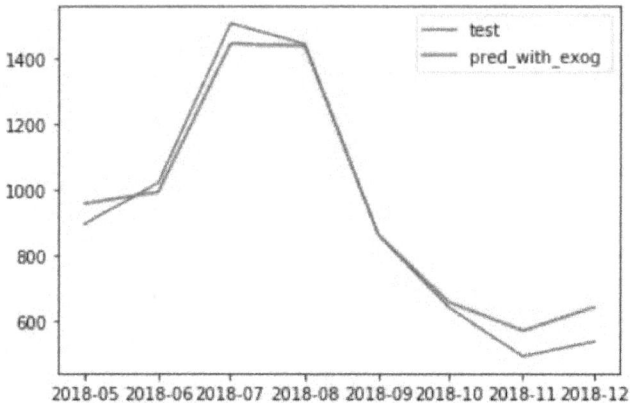

After all, it did not change so much.

I introduced the flow of time-series analysis. Since statsmodels is a really useful library, I hope you can use it to have a wonderful time-series analysis for your purpose.

Time-Series Visualization in Python

Clustering and Converting Time-Series Data with Python

Much of the real-world data is information that changes over time or time-series data. If you look around us, you can easily find data such as behavior data and power consumption data of users who visit your website.

This time, let's examine how to use the "clustering" method to find patterns that repeatedly appear from "time-series data."

Data Used

In recent years, stories like renewable energy and electricity liberalization have been heard. This trend is not only the story of the USA, UK, or Japan, but even the famous Kaggle in the competition of machine learning and data analysis also prefers to use the content related to the power consumption.

Link:

https://www.kaggle.com/competitions?sortBy=rele
=all&search=electricity&page=1&pageSize=20

Here, we take advantage of this trend and use the following power consumption data sets among the data sets for machine learning provided by the UC Irvine Machine Learning Repository.

- Data download page URL
    - https://archive.ics.uci.edu/ml/datasets/Electricity LoadDiagrams20112014
- Characteristic
    - Data acquisition period: (2011/1/1, 2015/1/1)
    - No missing measurements
    - If usage starts or ends during the period, the measured value is displayed as 0
- Data form
    - CSV format
    - Column delimiter: semicolon (;)
    - Decimal separator: colon (,) (this is standard in Portugal)

Tools to Use

We use Python 3.6 and Jupyter Notebook as the data analysis environment. The tslearn library is used for clustering time-series data. The necessary tools are listed below, so install them as needed.

- Python (> 3.x) with Jupyter
- Library
  - pandas
  - scipy, numpy, cython (depends on tslearn)
  - tslearn

Acquisition of Data

```
$ wget -nv -O /tmp/LD2011_2014.txt.zip https://archive.ics.uci.edu/ml/machine-learning-databases/00321/LD2011_2014.txt.zip
$ unzip /tmp/LD2011_2014.txt.zip -d /tmp/LD2011_2014
$ ls -l -h /tmp/LD2011_2014
total 679M
-rw-r----- 1 hccho hccho 679M Mar  3  2015 LD2011_2014.txt
drwxrwxr-x 2 hccho hccho 4.0K Mar 17  2015 _MACOSX
```

Looking at the contents of the text, the first column shows time information at 15-minute intervals (no column name), and after that, the power consumption of the client (MT_XXX) is recorded. In the example below, the width is 100 characters, but it is actually longer.

You can see that the data columns are divided by semicolons (';'). The data is reported with one line for each day of the period. The data has missing values; you can note, for example, that for Meter 1 and 2, there is no data.

```
$ head -n 10 /tmp/LD2011_2014/LD2011_2014.txt | cut -c 1-100
"";"MT_001";"MT_002";"MT_003";"MT_004";"MT_005";"MT_006";"MT_007";"MT_008";"MT_009";"MT_010";"MT_011
"2011-01-01 00:15:00";0;0;0;0;0;0;0;0;0;0;0;0;0;0;0;0;0;0;0;0;0;0;0;0;0;0;0;0;0;0;0;0;0;0;0;0;0;0;0;0;
"2011-01-01 00:30:00";0;0;0;0;0;0;0;0;0;0;0;0;0;0;0;0;0;0;0;0;0;0;0;0;0;0;0;0;0;0;0;0;0;0;0;0;0;0;0;0;
"2011-01-01 00:45:00";0;0;0;0;0;0;0;0;0;0;0;0;0;0;0;0;0;0;0;0;0;0;0;0;0;0;0;0;0;0;0;0;0;0;0;0;0;0;0;0;
"2011-01-01 01:00:00";0;0;0;0;0;0;0;0;0;0;0;0;0;0;0;0;0;0;0;0;0;0;0;0;0;0;0;0;0;0;0;0;0;0;0;0;0;0;0;0;
"2011-01-01 01:15:00";0;0;0;0;0;0;0;0;0;0;0;0;0;0;0;0;0;0;0;0;0;0;0;0;0;0;0;0;0;0;0;0;0;0;0;0;0;0;0;0;
"2011-01-01 01:30:00";0;0;0;0;0;0;0;0;0;0;0;0;0;0;0;0;0;0;0;0;0;0;0;0;0;0;0;0;0;0;0;0;0;0;0;0;0;0;0;0;
"2011-01-01 01:45:00";0;0;0;0;0;0;0;0;0;0;0;0;0;0;0;0;0;0;0;0;0;0;0;0;0;0;0;0;0;0;0;0;0;0;0;0;0;0;0;0;
"2011-01-01 02:00:00";0;0;0;0;0;0;0;0;0;0;0;0;0;0;0;0;0;0;0;0;0;0;0;0;0;0;0;0;0;0;0;0;0;0;0;0;0;0;0;0;
"2011-01-01 02:15:00";0;0;0;0;0;0;0;0;0;0;0;0;0;0;0;0;0;0;0;0;0;0;0;0;0;0;0;0;0;0;0;0;0;0;0;0;0;0;0;0;
$ tail -n 10 /tmp/LD2011_2014/LD2011_2014.txt | cut -c 1-100
"2014-12-31 21:45:00";2,53807106598985;22,0483641536273;1,73761946133797;156,50406504065;86,58536585
"2014-12-31 22:00:00";1,26903553299492;22,0483641536273;1,73761946133797;164,634146341463;93,9024390
"2014-12-31 22:15:00";2,53807106598985;22,0483641536273;1,73761946133797;160,569105691057;87,8048780
"2014-12-31 22:30:00";2,53807106598985;22,0483641536273;1,73761946133797;162,60162601626;80,48780487
"2014-12-31 22:45:00";1,26903553299492;22,0483641536273;1,73761946133797;156,50406504065;85,36585365
"2014-12-31 23:00:00";2,53807106598985;22,0483641536273;1,73761946133797;150,406504065041;85,3658536
"2014-12-31 23:15:00";2,53807106598985;21,3371266002845;1,73761946133797;166,666666666667;81,7073170
"2014-12-31 23:30:00";2,53807106598985;20,6258890469417;1,73761946133797;162,60162601626;82,92682926
"2014-12-31 23:45:00";1,26903553299492;21,3371266002845;1,73761946133797;166,666666666667;85,3658536
"2015-01-01 00:00:00";2,53807106598985;19,9146514935989;1,73761946133797;178,861788617886;84,1463414
```

Reading and Checking Raw Data

```python
import logging
logging.basicConfig()
logger = logging.getLogger()
logger.setLevel(logging.WARN)
import itertools
import calendar
from collections import Counter
from datetime import datetime, timedelta
import pandas as pd
import numpy as np
pd.set_option('max_rows', 20)
from tslearn.utils import to_time_series_dataset
from tslearn.clustering import TimeSeriesKMeans
from tslearn.preprocessing import TimeSeriesScalerMeanVariance, TimeSeriesScalerMinMax
import matplotlib.pyplot as plt
import matplotlib.markers as plt_markers
%matplotlib notebook
```

Also, although a bit long, define the following classes for data visualization.

```python
class Visualizer (object):
    def __init__ (self):
        self._markers = itertools.cycle ([
            'o', 'v', '^', '<', '>', '1', '2', '3', '4', '8',
            's', 'p', '*', 'h', 'H', '+', 'x', 'D', 'd', 'P'])
        self._colors = itertools.cycle (['b', 'g', 'r', 'c', 'm', 'y'])
        self._name_to_marker = {}
    def _make_xticks_and_xlabels (self, xs_ticks, xtick_step, xs_labels):
        """ Generate xtick and xlabel of a specific unit to display x-axis label
        """
        shown_xticks = []
        shown_xlabels = []
        for xtick, xlabel in zip (xs_ticks, xs_labels):
            check_digit = 0
            check_digit |= 1 if xlabel.minute == 0 else 0
            check_digit |= 2 if xlabel.hour == 0 else 0
            check_digit |= 4 if xlabel.day == 1 else 0
            if (xtick_step == 'h' and (check_digit & 1) == 1) \
                or (xtick_step == 'd' and (check_digit & 3) == 3) \
                or (xtick_step == 'm' and (check_digit & 7) == 7):
                shown_xticks.append (xtick)
                shown_xlabels.append (
                    '{} ({})'. format (str (xlabel), calendar.day_abbr [xlabel.wee
                    logger.info ('Adding xtick, xlabel: {}, {}'. format (xtick, xlabe
        return shown_xticks, shown_xlabels
    def draw_graphs (self, df, cnt = None, scatter = True,
                     xlabel = 'Time', xtick_step = 'm',
                     ylabel = 'kW / 15-min', ylog = False,
                     figsize = (12,8), dotsize = 3, alpha = 0.3):
        """ Visualization function for time series data.

        df: Pandas DataFrame. The `dt` column corresponds to the x axis, and one
corresponds to the y axis (one graph).
        cnt: Information to be displayed in addition to the legend name for each
number of samples belonging to the cluster).
        scatter: Use scatter graph for True and line plot graph for False.
        xlabel: x-axis name.
        xtick_step: h (hour), d (day), m (month).
        ylabel: The y-axis name.
        ylog: If True, set the y axis to log scale.
        figsize: Canvas size.
        dotsize: If scatter = True, set the size of each point.
        alpha: Set the transparency of the graph.
        """
        plt.subplots (1, figsize = figsize)
        # y-axis log scale?
        if ylog:
            plt.yscale ('log')
        # Draw a graph
        xs_labels = df ['dt']
        xs_ticks = np.arange (len (xs_labels))
        ys_columns = df.columns.tolist ()
        ys_columns.remove ('dt')
        for ys_col in ys_columns:
            ys = df [ys_col]
            # Save the graph name and its marker / color information
            if ys_col not in self._name_to_marker:
                self._name_to_marker [ys_col] = {
                    'marker': next (self._markers),
                    'color': next (self._colors)}
            # If cnt = None, add additional information to the name of each graph
            if cnt:
                ys_label = '{}: {}'. format (ys_col, cnt [ys_col])
            else:
                ys_label = ys_col
            # Draw a scatter / line graph.
            if scatter:
                plt.scatter (
                    xs_ticks, ys,
                    marker = self._name_to_marker [ys_col] ['marker'],
                    c = self._name_to_marker [ys_col] ['color'],
                    s = dotsize, alpha = alpha,
                    label = ys_label)
            else:
                plt.plot (
```

Then, read the above data with Python because it is a CSV file; it can be easily read using Pandas. However, since the delimiter is customized, let's specify the option as follows and load it.

```python
filepath = '/tmp/LD2011_2014/LD2011_2014.txt'
# Load as Pandas DataFrame.
df_raw = pd.read_csv (filepath, sep = ';', decimal = ',')
# Give "str_dt" as the column name of the measurement time column, and give it to Python datetime
# Add a new "dt" column converted to the format.
df_raw = df_raw.rename (columns = {'Unnamed: 0': 'str_dt'})
df_raw ['dt'] = df_raw ['str_dt'].apply (
    lambda s: datetime.strptime (s, '%Y-%m-%d %H:%M:%S'))
# Verify the loading result.
# Try to output only a part because there are many columns.
df_raw [['str_dt', 'dt', 'MT_001', 'MT_002', 'MT_003', 'MT_004', 'MT_005']]
```

145

	str_dt	dt	MT_001	MT_002	MT_003	MT_004	MT_005
0	2011-01-01 00:15:00	2011-01-01 00:15:00	0.000000	0.000000	0.000000	0.000000	0.000000
1	2011-01-01 00:30:00	2011-01-01 00:30:00	0.000000	0.000000	0.000000	0.000000	0.000000
2	2011-01-01 00:45:00	2011-01-01 00:45:00	0.000000	0.000000	0.000000	0.000000	0.000000
3	2011-01-01 01:00:00	2011-01-01 01:00:00	0.000000	0.000000	0.000000	0.000000	0.000000
4	2011-01-01 01:15:00	2011-01-01 01:15:00	0.000000	0.000000	0.000000	0.000000	0.000000
5	2011-01-01 01:30:00	2011-01-01 01:30:00	0.000000	0.000000	0.000000	0.000000	0.000000
6	2011-01-01 01:45:00	2011-01-01 01:45:00	0.000000	0.000000	0.000000	0.000000	0.000000
7	2011-01-01 02:00:00	2011-01-01 02:00:00	0.000000	0.000000	0.000000	0.000000	0.000000
8	2011-01-01 02:15:00	2011-01-01 02:15:00	0.000000	0.000000	0.000000	0.000000	0.000000
9	2011-01-01 02:30:00	2011-01-01 02:30:00	0.000000	0.000000	0.000000	0.000000	0.000000
...	...	...	...	...	...	...	...
140246	2014-12-31 21:45:00	2014-12-31 21:45:00	2.538071	22.048364	1.737619	156.504065	86.585366
140247	2014-12-31 22:00:00	2014-12-31 22:00:00	1.269036	22.048364	1.737619	164.634146	93.902439
140248	2014-12-31 22:15:00	2014-12-31 22:15:00	2.538071	22.048364	1.737619	160.569106	87.804878
140249	2014-12-31 22:30:00	2014-12-31 22:30:00	2.538071	22.048364	1.737619	162.601626	80.487805
140250	2014-12-31 22:45:00	2014-12-31 22:45:00	1.269036	22.048364	1.737619	156.504065	85.365854
140251	2014-12-31 23:00:00	2014-12-31 23:00:00	2.538071	22.048364	1.737619	150.406504	85.365854
140252	2014-12-31 23:15:00	2014-12-31 23:15:00	2.538071	21.337127	1.737619	166.666667	81.707317
140253	2014-12-31 23:30:00	2014-12-31 23:30:00	2.538071	20.625889	1.737619	162.601626	82.926829
140254	2014-12-31 23:45:00	2014-12-31 23:45:00	1.269036	21.337127	1.737619	166.666667	85.365854
140255	2015-01-01 00:00:00	2015-01-01 00:00:00	2.538071	19.914651	1.737619	178.861789	84.146341

140256 rows × 7 columns

The visualization of these five clients (MT_001 ~ 005) is as follows. In both cases, the data before 2012 is 0, indicating that no contract of consuming electricity was made.

```
viz.draw_graphs(
    df_raw[['dt', 'MT_001', 'MT_002', 'MT_003', 'MT_004', 'MT_005']],
    xtick_step='m',
    dotsize=2,
    alpha=0.4)
```

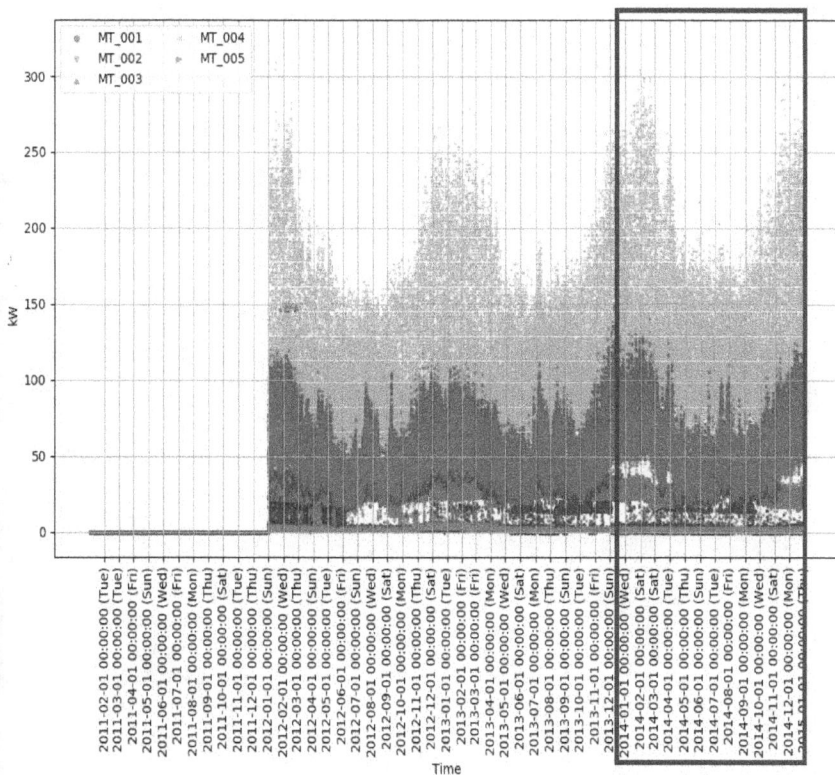

In this case, we will use only the data for the year 2014 (the part surrounded by a brown square) to speed up the experiment.

```
df_sample = df_raw[df_raw['dt'] >= '2014-01-01 00:00:00']
```

Clustering of Time-Series Data and Distance Function

One thing to keep in mind when handling time-series data is the choice of a method to calculate the distance (similarity) between two time-series data. Depending on the task, you may want to recognize the same pattern even if the data has some deviation from the time axis.

For example, in a speech recognition task, the speed of the same utterance varies from speaker to speaker (or even from the same speaker, depending on the situation). A distance function such as Dynamic Time Wrapping (DTW) is used to fill for such a gap.

However, in the power consumption data used this time, deviation from the time axis means an important difference (before and after work or between day and night), so Euclidean distance is used instead of DTW.

Euclidean distance algorithm is the application of a mathematical formula that is applied to the study of graphs and is extremely useful for discovering the similarity between two sets of data. Within the functionalities of the Euclidean distance algorithm, we can determine the similarity between two things or pairs of data that stand out. With it, we can get data schemas that identify elements that have similar characteristics.

Before clustering, convert the power consumption data of each client to a one-dimensional vector representation. This is the input data of the tslearn library used this time. Also, if all data values are 0, they are excluded as exceptions.

The following code will convert the data of each individual client into one dimensional vector representation.

```python
def get_target_column_names (cols):
    columns = cols.tolist ()
    columns.remove ('str_dt')
    columns.remove ('dt')
    return columns
def remove_all_zero_ys (yss):
    filtered_yss = []
    for ys in yss:
        all_zero = True
        for y in ys:
            if y [0]! = 0.0:
                all_zero = False
                break
        if not all_zero:
            filtered_yss.append (ys)
    return np.array (filtered_yss)
# X-axis data.
xs = df_sample ['dt']
# Y-axis data (multiple).
y_columns = get_target_column_names (df_sample.columns)
yss = to_time_series_dataset (
    df_sample.as_matrix (y_columns) .T)
# Delete all zero data
filtered_yss = remove_all_zero_ys (yss)
```

Next, cluster the data with a large number of classes as 12 (this time, we did not find the optimal number of clusters, but we actually need to do this by trial and error). At this time, fix the random value used when initializing the cluster center for the reproducibility of the experiment.

A random seed determines the starting point when a computer generates a series of random numbers. It can be a random number, but it usually comes from the seconds of a computer system's clock.

Random numbers or data generated by the Python random module are not really random, but pseudo-random (i.e., PRNG), i.e., deterministic. It generates numbers from a certain value. This value is nothing more than a starting value (seed value), i.e., the random module uses the starting value as a basis to generate a random number.

In general, the seed value is the previous number produced by the generator. However, when using the random number generator for the first time, there is no previous value. Therefore, the current system time is used as the starting value.

```
n_clusters            = 12
clustering_metric     = 'euclidean'
rand_seed             = 12345
km = TimeSeriesKMeans(
    n_clusters=n_clusters,
    metric=clustering_metric,
    random_state=rand_seed)
km.fit(filtered_yss)
```

Let's visualize the clustered result.

```
# Create a DataFrame with columns of `dt`,` cluster-0`, `cluster-1`, ...
data = []
for cluster_x in km.cluster_centers_:
    data.append (
        [point_x [0] for point_x in cluster_x]
    )
columns = ['cluster-{}'. format (idx) for idx in range (n_clusters)]
clusters = np.array (data) .T
df_clusters = pd.DataFrame (clusters, columns = columns)
df_clusters ['dt'] = xs.values
# Calculate the number of data belonging to the cluster.
cnt = Counter (km.labels_)
cluster_labels = {}
for k in cnt:
    cluster_labels ['cluster-{}'. format (k)] = cnt [k]
# Display graph.
viz.draw_graphs (df_clusters, cnt = cluster_labels,
        xtick_step = 'm',
        alpha = 0.4)
```

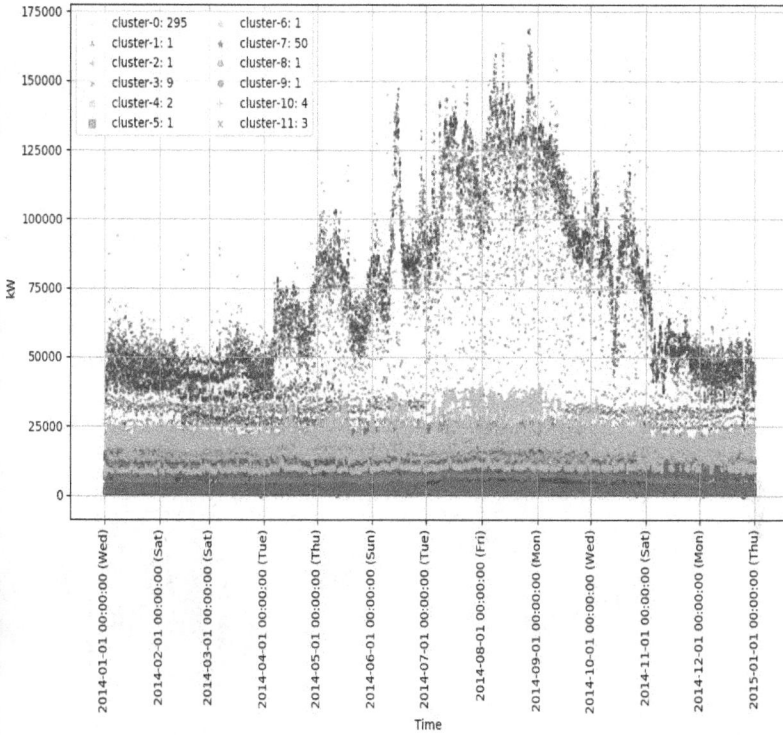

The following graph is an enlarged version of the data for the first 30 days in the above graph. The following characteristics appear from these two graphs.

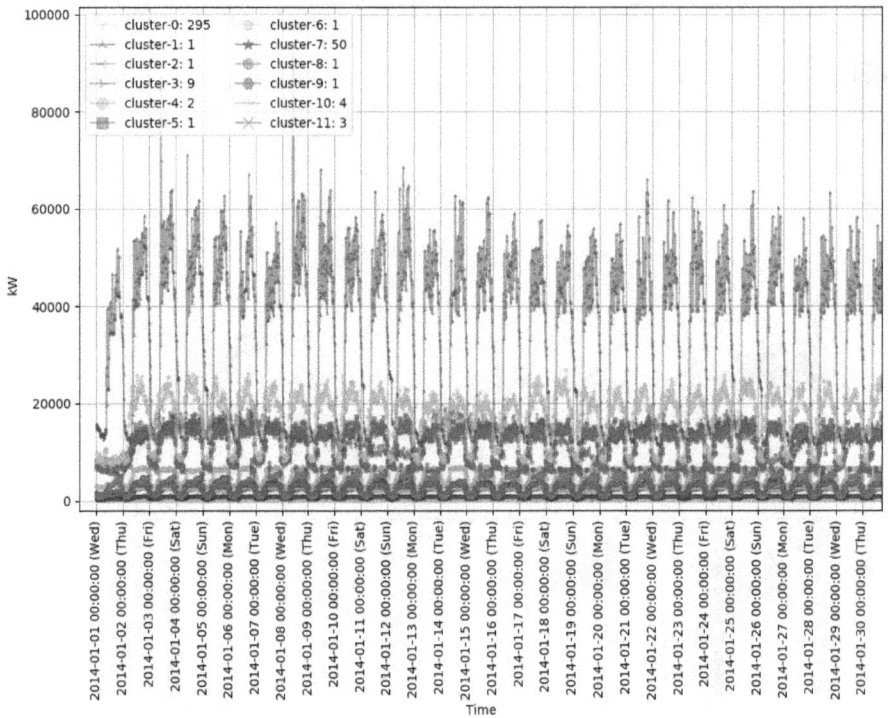

Legend:
- cluster-0: 295
- cluster-1: 1
- cluster-2: 1
- cluster-3: 9
- cluster-4: 2
- cluster-5: 1
- cluster-6: 1
- cluster-7: 50
- cluster-8: 1
- cluster-9: 1
- cluster-10: 4
- cluster-11: 3

1. The absolute value is very high, and the cluster that peaks in summer (cluster-1) is prominent.

2. There are only half clusters containing a plurality of time-series data. Also, looking at the second graph, in many cases (cluster-1, 2, 5, 6) in which one cluster was generated from one data were much higher in absolute value (10,000 ~ 70,000kW or more).

Clients belonging to cluster-1 require up to 40,000 kW of power consumption in summer, which is about 13,000 times the maximum power consumption of ordinary households (100 V x 30 A) of 3 kW. Considering such power consumption, it is likely that this is a large-scale factory, and in summer, it may need to run coolers to keep the inside temperature normal.

Then, if you plot only cluster-2, 5, and 6 excluding cluster-1, whose absolute value is abnormally high, you will get the following graph.

```
# Plot three large clusters (cluster-2, 5, 6) excluding cluster-1
target_clusters = ['cluster-{}'.format(idx) for idx in [2,5,6]]
filtered_df_clusters = df_clusters [['dt'] + target_clusters]
viz.draw_graphs (filtered_df_clusters, cnt = cluster_labels,
            xtick_step = 'd',
            scatter = False,
            alpha = 0.4)
```

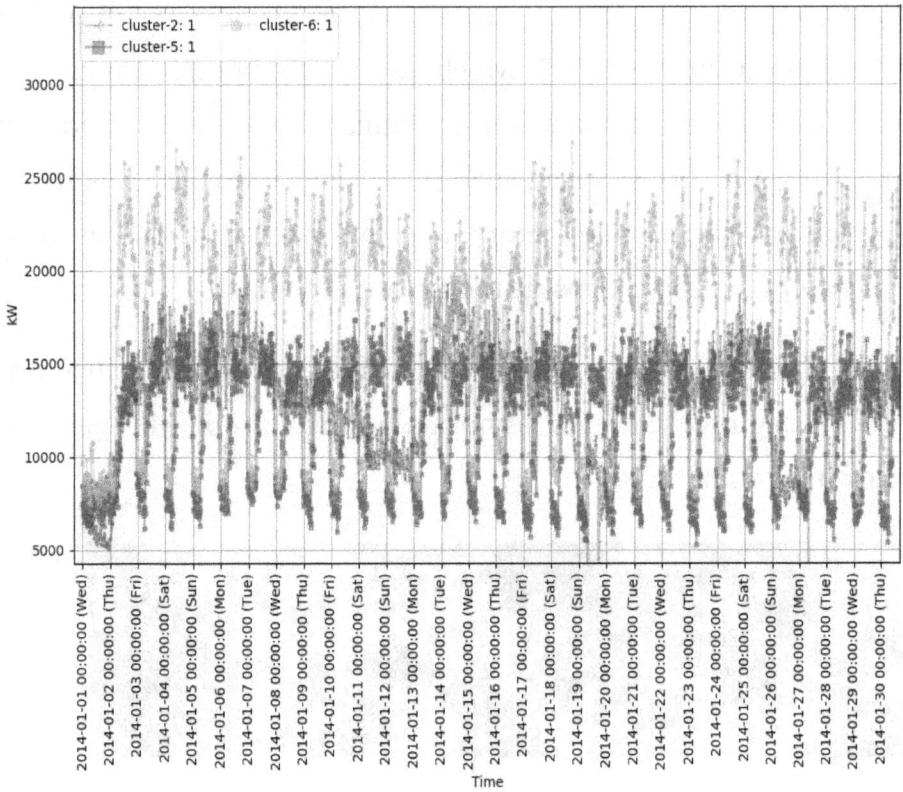

First, you can see that the power consumption is low on January 1 because it is closed. And although cluster-1 and 5 have the difference of the maximum power consumption, it turns out that the pattern is very similar. It can be guessed that they are in the same industry with different scales. However, the power consumption of clients belonging to cluster-2 has a characteristic that falls during the weekend (Saturday and Sunday). It seems that the data is for places that do not open on weekends, such as public institutions.

Next, this time we will look at the characteristics of four clusters (cluster-0, 3, 7, 10) with small absolute values of power consumption. This cluster contains a lot of data. The number of data belonging to the group is on the right side of the cluster name. The first graph is for one year and the second graph is for the first 30 days. The most prominent feature in this graph is that the cluster-0 and 7 show a very stable pattern.

Considering the power consumption, it seems that it is an ordinary household. In addition, cluster-10 has the characteristic of reducing power consumption over the weekend. Although the absolute value is 6-7 times different, it is considered that it is the same industry as cluster-2 seen before.

Let's take a look at the four clusters (cluster-4, 8, 9, 11) where the last power consumption was intermediate. Here I would like to show cluster-8 and 9 and cluster4 and 11 in different graphs. Let's first look at the graphs for cluster-8 and 9. The graphs of cluster-8 and 9 both have large differences between the minimum and maximum values of power consumption and are similar to each other in absolute value.

The fact that it was recognized as another cluster means that the maximum power consumption in February to March in the time-series data belonging to cluster-9 was less than half of the other time, probably recognized as a cluster.

However, there is not enough difference in terms of result between cluster 8 and 9 as their trend is very similar, and it indicates that they are the members of the same group perhaps in the commercial sector with approximately the same size.

Link:

https://www.eia.gov/totalenergy/data/monthly/r

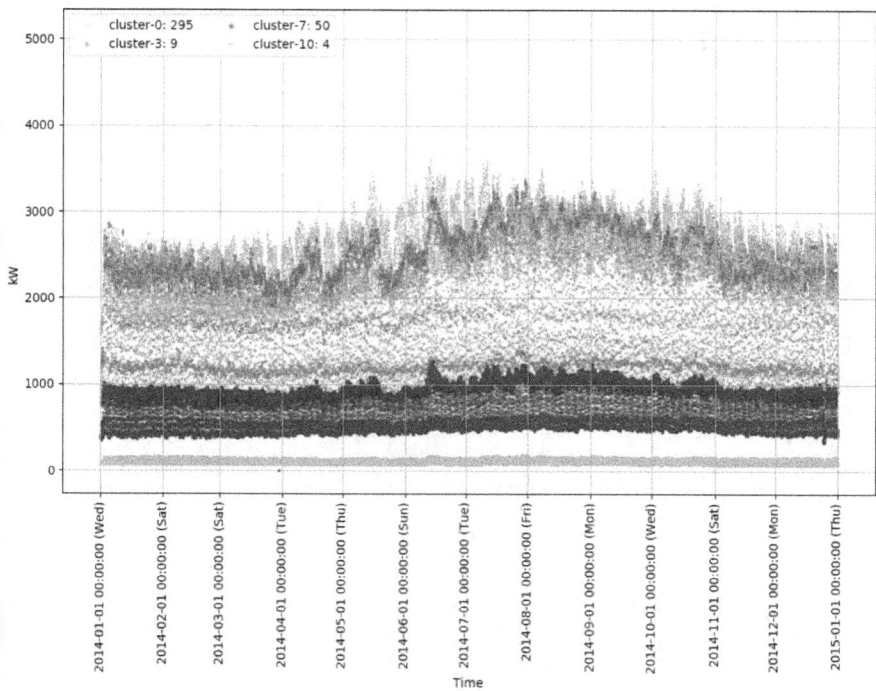

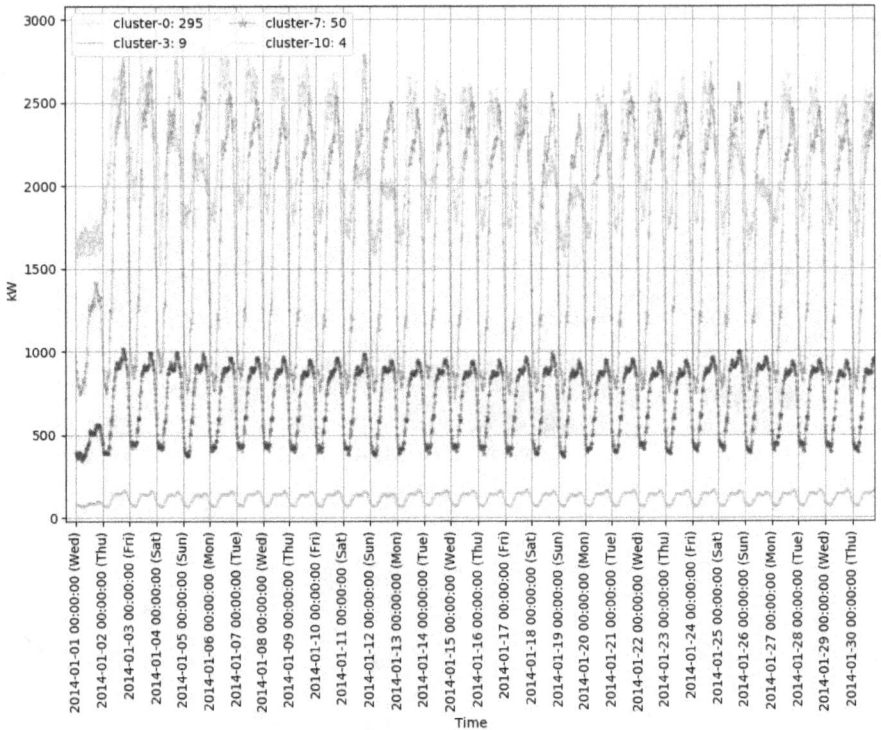

But what types of industries have such extreme differences in power consumption? Continuing, comparing cluster-4 with 11, since it is very beautifully divided, the patterns tend to be very similar, so, likely, they are also clients of the same industry but of different sizes.

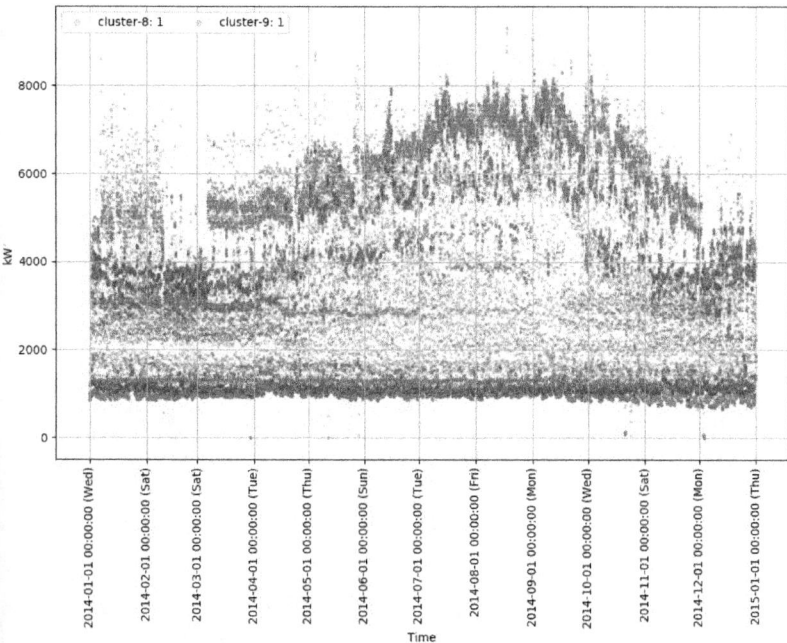

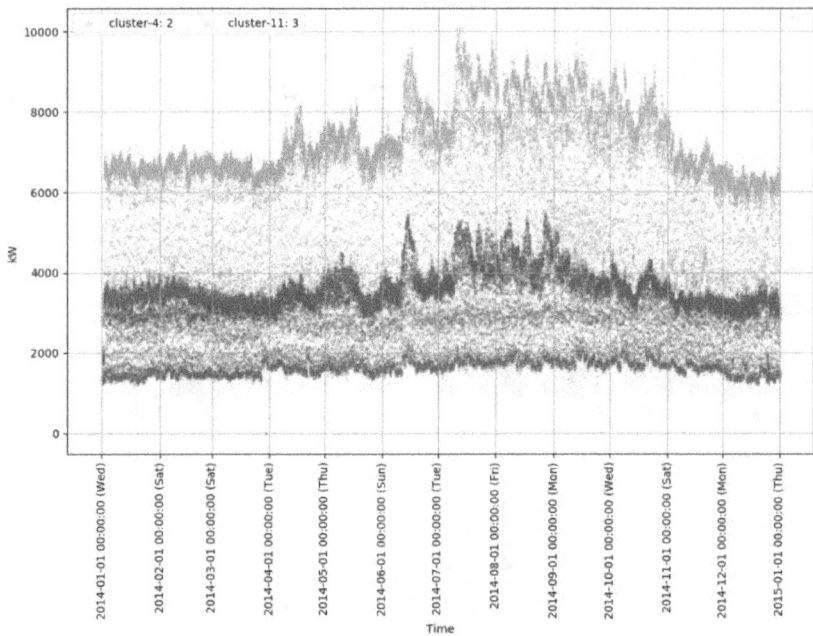

In this section, we examined the frequent time-series patterns by applying a clustering method to power consumption data, which is a typical time-series data.

Since the data consisted of only one feature, the power consumption per unit time, I thought at first that it would be easy to get a visible result, but in practice, there were some parts that to you may feel somewhat complex to analyze. But do not worry, just move through the following content of the book, and you will get smoother with practice in time-series analysis with Python.

Matplotlib and NetworkX

Now in this section, I will show the use of Python libraries matplotlib (which serves to produce scientific drawings) and networkx (which helps to manipulate and draw complex networks).

As a first example, I will show the code to produce the first figure of this chapter below, which in turn shows a representation of the time-series [2, 3, 10, 2, 4, 5] using matplotlib.

```
import matplotlib . pyplot as plt
y =[2 ,3 ,10 ,2 ,4 ,5]
plt . grid ( True )
plt . plot ( y )
plt . show ()
```

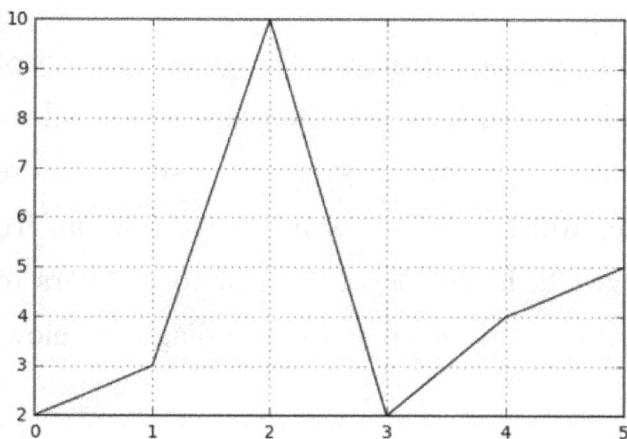

Python has the necessary data structures, such as lists and integers. For example, the second line of the first code example assigns the variable and the list of integers [2,3,10,2,4,5], which in this case, we are interpreting as a time-series. In Python, many libraries extend the capacity of this program. As an example, I can mention the free matplotlib (https://matplotlib.org), which is mentioned in the first line of the first example.

The program has a syntax similar to that of MATLAB™. The next two lines modify a graphic object. With plt.grid(True), a grid is superimposed on the graphic. With plt.plot(y), the lines corresponding to the time-series given by the list we have denoted with y are added. Finally, with plt.show(), the drawing is shown.

Although Python is a somewhat complicated language in terms of complex scientific calculation, it is not necessary to use all its complexity to start using it. I am going to give a slightly more complicated example, which will be useful from now on. To define the visibility graph, it will be convenient to use bars to indicate the time-series values, as shown in the diagram below, the code snippet.

```
import matplotlib.pyplot as plt
y=[2,3,10,2,4,5]
plt.grid(True)
plt.bar(range(len(y)),
        y,
        width=0.2,
        align='center')
plt.show()
```

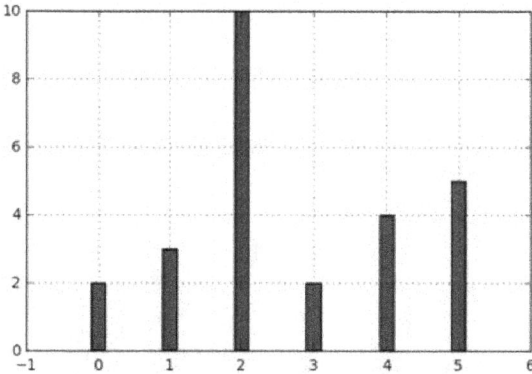

Graphs – Nodes in Python

The next figure shows a drawing of graph P, which we ended in the previous paragraph made with Python and the libraries matplotlib and networkx (the networkx website is located at https://networkx.github.io/). As we can see in the code, shown below, after declaring that we want to use networkx with the import instruction networkx as nx, we can use the functions nx.Graph to create a chart and functions that modify the different elements of a graph (vertices, edges, and labels of the vertices), to create a drawing like before we need to use plt.show() every time.

Note that, in this case, we have created a graph using the nx.Graph function, giving as an argument to the list of edges of the graph to be considered.

```
import matplotlib.pyplot as plt
import networkx as nx
P=nx.Graph([('a','b'),('b','c')])
pos=nx.spectral_layout(P)
nx.draw_networkx_nodes(P,pos,node_size=800)
nx.draw_networkx_edges(P,pos)
nx.draw_networkx_labels(P,pos,font_size=25)
cut = 0.07
plt.xlim(-cut,1+cut)
plt.ylim(-cut,1-cut)
```

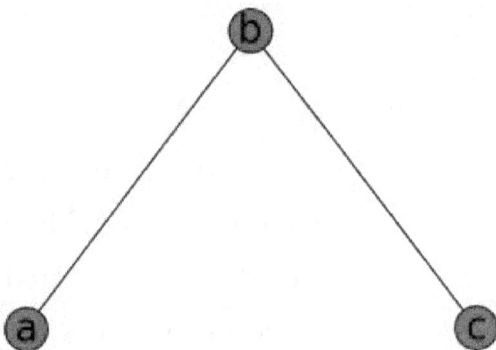

Once the graph has been created in Python, it is possible to use the computer to obtain graph properties. For example:

P. nodes () produces:

['a', 'c', 'b']

P. degree('b') produces:

2.

Application to the Visibility Graph

Visibility Graph

In this article (Lacasa and others, 2008), the authors introduce the visibility graph associated with a time-series as a tool for analyzing different properties of the series, using the techniques and terminology of graph theory. On the other hand, they also observe that the visibility graph remains invariant under certain non-essential changes of the time-series, such as translation or rescaling.

Article Link:

https://www.researchgate.net/publication/3012328

me series to complex networks The visibility g

The visibility graph has its vertices, i.e., the data of the time-series, the vertices can be obtained as V = 0, 1, 2, n 1. The data are declared adjacent to each other in such a way that, in the time-series drawing, the upper parts of their corresponding bars are "visible" to each other, considering the bars as "pairs." Examples are shown in the next two Figures.

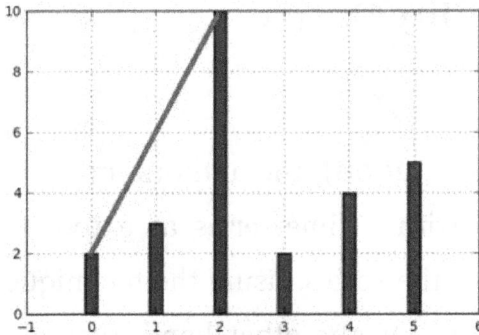

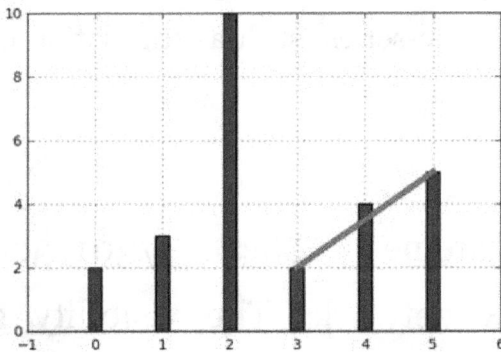

It is then immediate to determine the complete list of the edges of the visibility graph of a time-series, for example, [2, 3, 10, 2, 4, 5], and using networkx to draw a picture of the graph. The drawing is shown in the next figure.

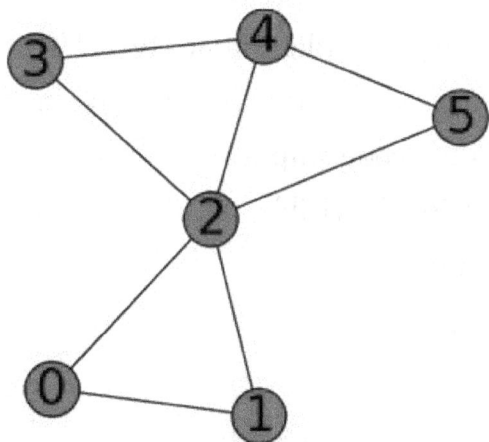

In this case, since the time-series considered has little data, it is feasible to explicitly list the edges of the visibility graph by simply using the time-series drawing. In the following sections, we will show how Python can help us to study more complicated time-series analysis.

The Formal Definition of the Visibility Graph

Determining whether two data are adjacent in the visibility graph in a precise way is an exercise in analytical geometry. For example, (Lacasa and others, 2008) described that the data (t_a, y_a), (t_b, y_b) are declared adjacent in the visibility graph and showed that for all t_c with $t_a < t_c < t_b$ is fulfilled:

$$y_c < y_b + (y_a - y_b) \frac{t_b - t_c}{t_b - t_a}.$$

In Python, we can define a function that determines two different data in a given time-series, and such data are adjacent in the time-series visibility graph. In the code snippet shown below, we define this function. The function is_visible returns true if the data a, b are adjacent and if not false.

```
def is_visible(y,a,b):
    isit = True
    c = a+1
    while isit and c < b:
        isit = y[c]<y[b]+(y[a]-y[b])*((b-c)/float(b-a))
        c = c+1
    return isit
```

On the other hand, the next image shows the code to finish the visibility graph of a time-series, using the function is_visible from the previous code. In this case, a variable called eds. is used to collect the edges in a list. For each data, *a* in the time-series each *b* that is greater than *a* is determined if *b* is visible from *a*, and only it is the edge *(a,b)* added to the eds list. The visibility_graph function finally returns the visibility graph of the time-series ts.

```
def visibility_graph(ts):
    eds = []
    for a in range(len(ts)):
        for b in range(a+1,len(ts)):
            if is_visible(ts,a,b):
                eds.append((a,b))
    return nx.Graph(eds)
```

In the next section, we will apply this code to a much more complicated time-series analysis.

The Logistic Mapping

A simple way to analyze a time-series in a set of real numbers X is using the iterations of a function $f: X \to X$. To construct a time-series in such a way, an initial condition $x_0 \in X$ is taken, and the subsequent data is defined recursively for t > 0 as $x_t = f(x_{t-1})$.

One of the most studied cases, which we include here only as an example, is the logistic mapping, in which X = [0, 1] is taken, that means, the interval of real numbers between 0 and 1, and the function $f(x) = \mu x(1 - x)$, where μ satisfies $0 \le \mu \le 4$.

We refer the reader interested in the mapping properties (Devaney (1989), where it is shown that for absolute values of μ, the iterations show the phenomenon of chaos.

Link: https://bu.ac.bd/uploads/BUJ1V4I1/15.%2

The code below shows how to obtain the iterations of the logistic mapping to produce the next chart, where the time-series obtained with initial condition $x_0 = 0.3$, the value of parameter $\mu = 3.8$, and with 50 iterations is drawn.

```
import matplotlib.pyplot as plt
def f(x):
    return 3.8*x*(1-x)
vals=[0.3]
iteraciones=50
for i in range(iterations):
    new = vals[-1]
    vals.append(f(new))
plt.figure(figsize=(10,5))
plt.axis([-1,iterations+1,0,1])
plt.grid(True)
plt.bar(range(len(vals)),
        vals,
        width=0.2,
        align='center')
plt.show()
```

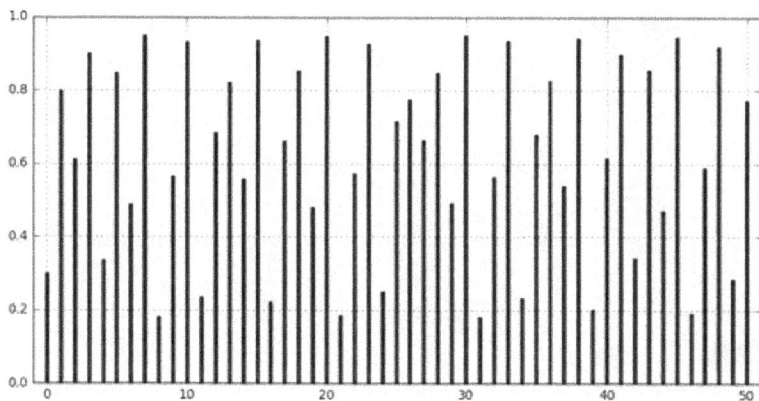

Note that in the code above, in addition to importing the library, a list is created, which contains the data of the time-series. Next, that list can be used to obtain the following figure, which shows the visibility graph of the time-series that originated with the iterations of the logistic mapping. In this case, we note that it is difficult to derive properties of the graph from the drawing alone due to the complexity of the drawing since it is not always clear whether an edge joins two vertices. Therefore, other tools are needed, such as the histogram of degrees, to study a complex graph.

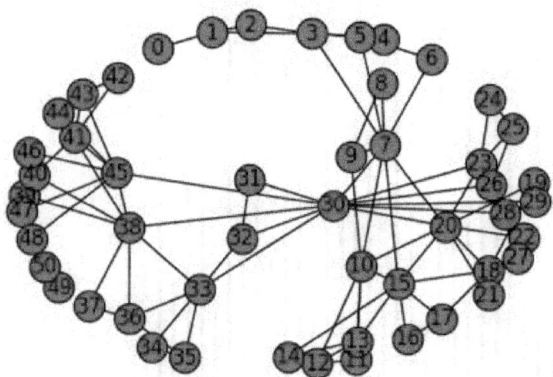

The following code can be used to produce a histogram that is shown in the next image representing the degrees of the previous graph.

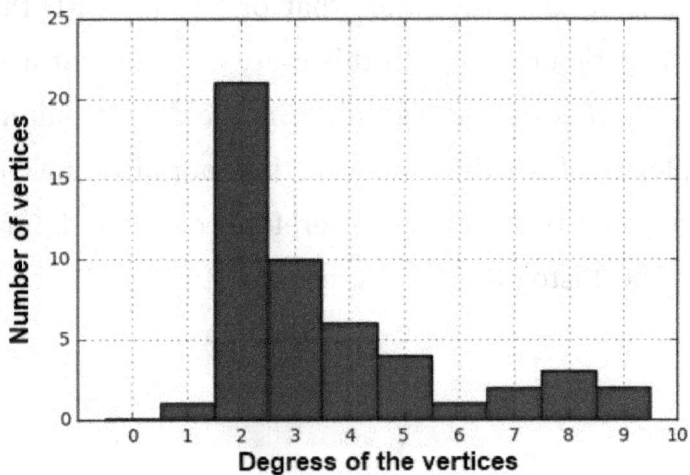

In literature, the distribution of the degrees of the vertices is the main tool for studying the graph of visibility, especially in cases where the graph has too many vertices. For example, the situation in which the time-series data are obtained at random from the chaotic case as we consider in this section.

In this part, I showed examples of using the Python program in the study of time-series analysis. There are two possible directions of research regarding the concepts explained here. One is theoretical and refers to studying the properties of the different visibility graphs that have been defined. For example, horizontal visibility graphs are characterized by combinatorial properties.

Another theoretical problem arises when considering that the time-series that can be studied in the computer are limited to having a finite number of terms. However, the concept of the visibility graph naturally extends to the infinite case. It leads to the study of infinite graph properties. For example, an interesting problem is considering how the convergence of a time-series is reflected in the graph.

Data Preprocessing and Time-Series

This part aims to demonstrate practically and understandably how to manipulate a series of historical data to bring out relevant insights. And, finally make a future forecast using machine learning techniques, in particular the ARIMA algorithm (autoregressive integrated moving average). This model is used frequently to make predictions on data where there is seasonality.

The ARIMA algorithm consists of a variation of an ARMA model with the addition of component d for stationarity, this model generates a series of future data using its own past data.

It requires three parameters to be used: $p,$ $d,$ and q.

The first parameter p refers to the number of lags used in the model, i.e., with what periodicity (seasonality) the model has to do. The second parameter d refers to the degree of differentiation, i.e., the number of times the data has had values subtracted in the past. At the same time, q represents the error in the form of a combination of past errors.

For this part, I used Python 3.7 in a Jupyter Notebook configured through the Miniconda package manager.

The Dataset

We start by importing the necessary libraries and datasets, after which the irrelevant columns will be removed from the dataset. Using pandas data, they are transformed into a format called DataFrame, that is, a sort of arrays.

```
# Import of the necessary libraries
import pandas as pd
import numpy as np
import matplotlib.pyplot as plt

plt.style.use('ggplot')


# Import the data by reading two .csv files via pandas
customers_df = pd.read_csv('customers_export.csv')
orders_df = pd.DataFrame(pd.Read_csv('orders_export.csv'))
cols_to_drop_cust = [0,1,3,4,5,6,7,9,10,12,16,17,18]
customers_df.drop(customers_df.columns[cols_to_drop_cust], axis=1, inplace=True)
```

By running these few lines of code and calling the command customers_df.head() we will have access to the first five lines:

	Email	Province Code	Zip	Accepts Marketing	Total Spent	Total Orders
0	@yahoo	BO	126	yes	25.0	1
1	NaN	RO	)19	no	0.0	1
2	@gmail	BO	136	yes	25 0	1
3	?gmail com	BO	128	yes	0.0	0
4	@hotmail	MB	71	yes	25.0	1

One thing that immediately caught my attention was seeing how more than half of the email addresses contained in this dataset have a number that often corresponds to the year of birth. So I decided to write a small algorithm to trace this number and possibly create a graph with the distribution of customers by age.

```
1. # I calculate the age based on the number extracted in the email
addresses
2. cust_age = customers_df. iloc [ :, 0 ] . str . extract ( '(\ d +)' ,
expand = False ) . dropna ( ) . astype ( int ) . tolist ( )
3. for n, i in enumerate ( cust_age ) :
4.         if 70 <= i <= 100 :
5.             cust_age [ n ] = i + 1900
6.     for x in reversed ( cust_age ) :
7.         if x> 2010 :
8.             cust_age. remove ( x )
9.     for x in reversed ( cust_age ) :
10.        if x < 70 :
11.            cust_age. remove ( x )
12.    for x in reversed ( cust_age ) :
13.        if x < 1900 :
14.            cust_age. remove ( x )
```

After doing a bit of cleaning and turning the list into an object using Pandas.Series I was able to create this chart, including only the first ten bars:

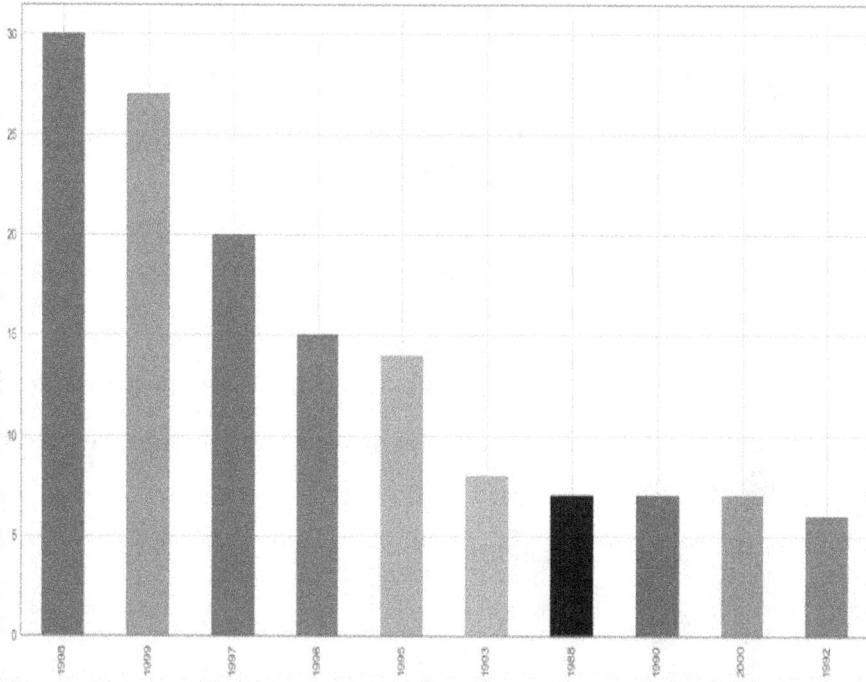

The next step is to analyze which province the customers order from, so just select the column for the region and using the method value_counts():

```
# Customers by region
provinces = pd. DataFrame ( customers_df. Iloc [ :, 1 ] )

# I delete rows containing 'null' data
provinces = provinces. dropna ( axis = 0 , how = 'all' )

provinces = provinces [ 'Province Code' ] . value_counts ( )
```

I get this 'fantastic' pie chart:

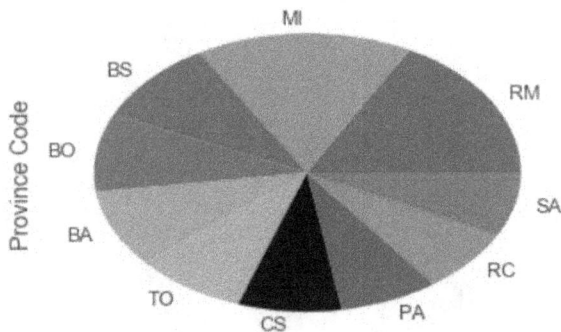

Another interesting point is to find out which are the most sold sizes, this could help to manage the warehouse in a better way, and in the future, it would allow using machine learning techniques to optimize the same.

```
#Convert list to df, drop NaN and plot
tshirt_sizes_df = pd. DataFrame ( tshirt_sizes )
tshirt_sizes_df = tshirt_sizes_df. dropna ( )

plt. plot ( pd. value_counts ( tshirt_sizes_df. values . flatten ( ) ) )
plt. xlabel ( 'Size' )
plt. ylabel ( 'Number of t-shirts' )
plt. title ( 'Number of t-shirts sorted by size' )
```

Number of T-shirt ordered per size

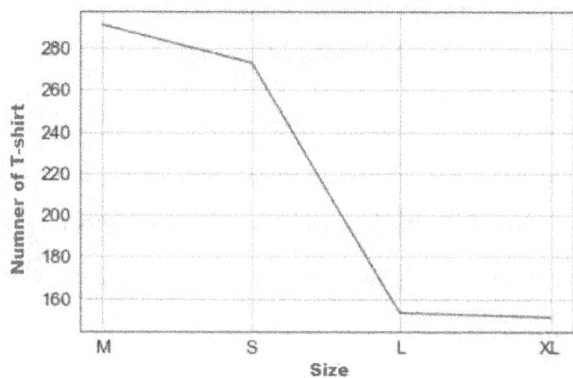

Time-Series Analysis

In this part, I will analyze the dataset based on the order dates, and the dataset contains information from 2016–05–01 to 2018–07–23, therefore more than two years. There are several time-series models and to evaluate which one should be used, it is necessary to use a toolkit that allows you to break down your series into its two components:

- Systematic: components containing consistent and usable trends in our model
- Non-systematic: components of the series that cannot be directly modeled

As we already know that a time-series consists of three systematic components: level, trend, seasonality, and an unsystematic one that is called as the noise.

Before you can use the model though, you need to do some order:

```
# column selection
created_at = pd. read_csv ( r 'C: \ Users \ roberto.sannazzaro \ Desktop \ sales_2016-05-
01_2018-07-23.csv' )

drops = [ 'Sale ID' , 'Order name' , 'Transaction type' , 'Sale type' , 'Sales channel' ,
'Shipping country' , 'Shipping region' , 'Product variant SKU' , 'POS location' , ' Billing
country ' ,
          'Billing region' , 'Billing city' , 'Net quantity' , 'Amount (before discount and
taxes)' , 'Line item discount' , 'Order discount' ,
          'Taxes' , 'Product type' , 'Product vendor' , 'Amount (after discounts before
tax)' , 'Shipping city' ]

created_at. drop ( drops, axis = 1 , inplace = True )
created_at = created_at. groupby ( created_at [ 'Date' ] ) [ 'Amount (after discounts and
taxes)' ] . sum ( ) . reset_index ( )
created_at = created_at. set_index ( 'Date' )
created_at. index = pd. to_datetime ( created_at. index )

y = created_at [ 'Amount (after discounts and taxes)' ] . resample ( 'M' ) . mean ( )

# I create the chart
y. plot ( figsize = ( 15 , 6 ) )
plt. xlabel ( 'Data' )
plt. ylabel ( 'Average sales in EUR' )
```

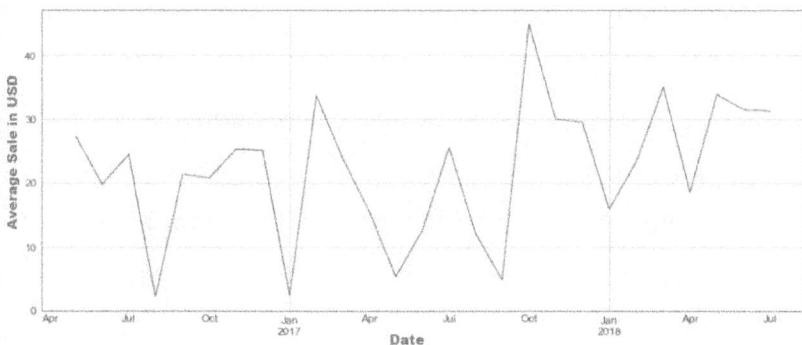

Projecting the series, it can be seen how the trend is slightly growing, a multiplicative breakdown is right for us in the case of a pattern with low growth/loss, or with a seasonal component can opt for an additive type model.

```
import statsmodels. api as sm
from pylab import rcParams
rcParams [ 'figure.figsize' ] = 20 , 12
decomposition = sm. tsa . seasonal_decompose ( y, model = 'multiplicative' , filt = None ,
two_sided = True , freq = 7 ) # freq = 1
fig = decomposition. plot ( )
plt. show ( )
```

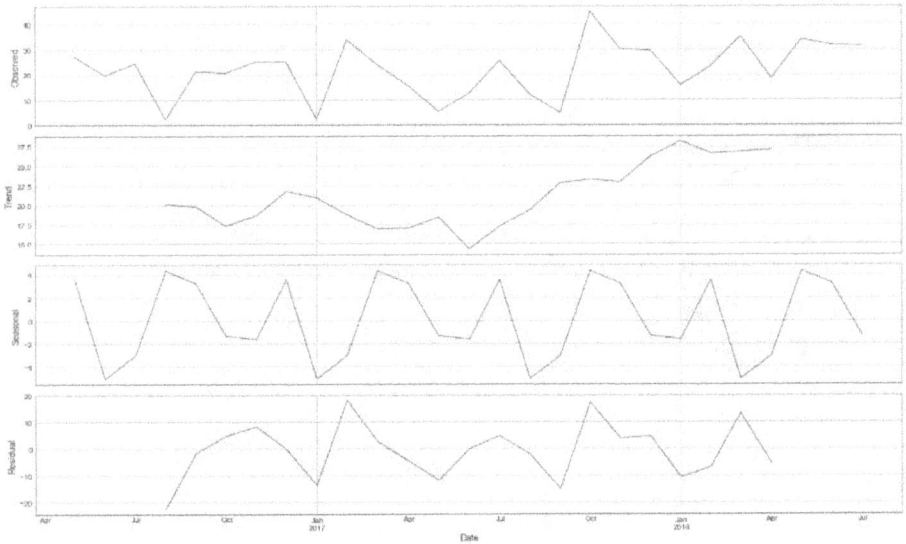

Thanks to this breakdown now, the seasonal component is clearly visible and can be observed as the growing trend. However, it is important to underline that in this case, being aware of the dynamics behind this commercial activity and being mindful of the fact that in the last year there were major changes in marketing and warehouse, I will opt to consider this trend as neutral, as I do not want the model to be influenced by these external factors.

The trend can be positive or negative, linear, or non-linear, so it is very important to know your dataset to understand when enough time has passed to identify a new trend.

At this moment, given the results of the breakdown, we can say that using an ARIMA model makes sense due to the seasonality. For this reason, now it is necessary to choose the parameters P, D, Q (seasonality, trend, noise) for the model.

One of the fastest and most effective systems for estimating these parameters is to carry out a search called "grid search" through different combinations of P, D, and Q values combined with a performance index.

For this, I will use the "Akaike information criterion" (AIC) like before.

The AIC, given a series of models, measures the quality of each in comparison with the other models generated by the combination of p, d, and q. The value of the AIC will allow understanding how each model takes the input data, taking into account the complexity of the model; therefore, the models that will have a better result (fit), using fewer features and that will provide a better (lower) score.

The library for Python pyramid-arima allows you to use this search system quickly, also generates an object with which you can directly call the method fit().

```
from pyramid. arima import auto_arima
stepwise_model = auto_arima ( y, start_p = 1 , start_q = 1 ,
                             max_p = 3 , max_q = 3 , m = 12 ,
                             start_P = 0 , seasonal = True ,
                             d = 1 , D = 1 , trace = True ,
                             error_action = 'ignore' ,
                             suppress_warnings = True ,
                             stepwise = True )
print ( stepwise_model. aic ( ) )
```

```
t ARIMA: order=(1, 1, 1) seasonal_order=(0, 1, 1, 12); AIC=nan, BIC=nan, Fit time=nan seconds
t ARIMA: order=(0, 1, 0) seasonal_order=(0, 1, 0, 12); AIC=126.466, BIC=127.744, Fit time=0.015 seconds
t ARIMA: order=(1, 1, 0) seasonal_order=(1, 1, 0, 12); AIC=112.105, BIC=114.661, Fit time=0.501 seconds
t ARIMA: order=(0, 1, 1) seasonal_order=(0, 1, 1, 12); AIC=nan, BIC=nan, Fit time=nan seconds
t ARIMA: order=(1, 1, 0) seasonal_order=(0, 1, 0, 12); AIC=121.439, BIC=123.356, Fit time=0.049 seconds
t ARIMA: order=(1, 1, 0) seasonal_order=(2, 1, 0, 12); AIC=nan, BIC=nan, Fit time=nan seconds
t ARIMA: order=(1, 1, 0) seasonal_order=(1, 1, 1, 12); AIC=nan, BIC=nan, Fit time=nan seconds
t ARIMA: order=(1, 1, 0) seasonal_order=(2, 1, 1, 12); AIC=nan, BIC=nan, Fit time=nan seconds
t ARIMA: order=(0, 1, 0) seasonal_order=(1, 1, 0, 12); AIC=124.337, BIC=126.255, Fit time=0.170 seconds
t ARIMA: order=(2, 1, 0) seasonal_order=(1, 1, 0, 12); AIC=117.850, BIC=121.045, Fit time=0.309 seconds
t ARIMA: order=(1, 1, 1) seasonal_order=(1, 1, 0, 12); AIC=117.579, BIC=120.775, Fit time=0.385 seconds
t ARIMA: order=(2, 1, 1) seasonal_order=(1, 1, 0, 12); AIC=119.568, BIC=123.402, Fit time=0.459 seconds
tal fit time: 1.908 seconds

112.10471050293648
```

According to Aike's test estimates, the ideal combination of variables scores 112.10 with a combination of parameters 1, 1, 0, 12.

To verify the effectiveness of the model, it statsmodel provides a method called plot_diagnostics():

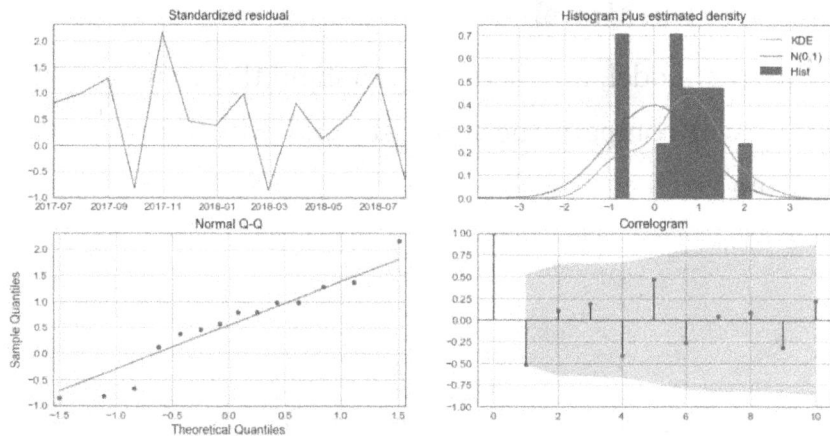

It is not perfect, as the line marked with KDE does not precisely follow an average distribution n (0, 1), this can also be seen in the qq graph: the distribution of residues does not differ much, for this reason; the forecasts will have an interval of greater confidence (less precise).

One of the main reasons why the model is not perfect is the lack of data, as the dataset contains information from the exact moment in which the commercial activity started, and therefore in an unstable situation and without any trend identifiable in the first 6 -8 months.

Validation of the Model

To validate the model, and better understand the accuracy of the forecasts, I will compare the forecast sales with the real ones, starting from 2017–12–31:

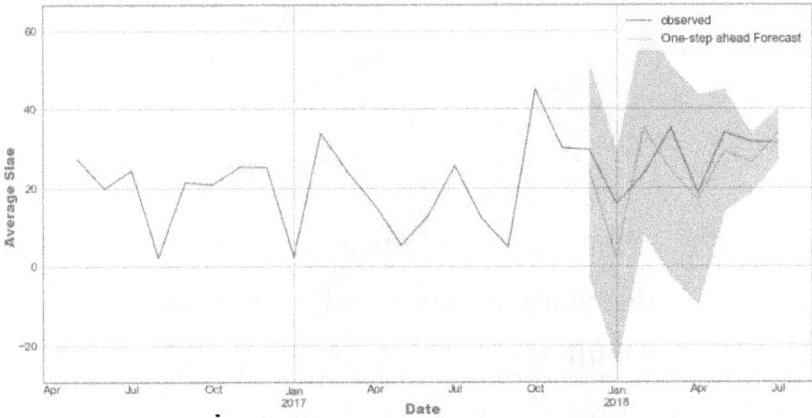

From what is possible to observe, the model predicts real data in the right way and can capture the seasonality presented in the data.

To have a more precise and measurable vision, it is necessary to calculate the R square coefficient, that is the square of the difference between estimated and real values, the more the R square value approaches 1, the better the forecasts will be. However, it is important to specify that a low R square value is not always associated with a low-quality model, as there is no range of standard values in which the R square coefficient must be found, but it depends on each case.

```
y_forecasted = pred. predicted_mean
y_truth = y [ '2017-01-01' : ]
mse = ( ( y_forecasted - y_truth ) ** 2 ) . mean ( )
print ( 'R sqaure: {}' . format ( round ( mse, 2 ) ) )
```

I would say not bad, even if it is not perfect.

As we have been able to verify, the model can capture the seasonality. However, the more the forecasts are pushed into the future, the more the confidence interval will increase. In this example, there is only one variable: the average of purchases per week. In contrast, the independent variable is described by the date and seasonality associated with it.

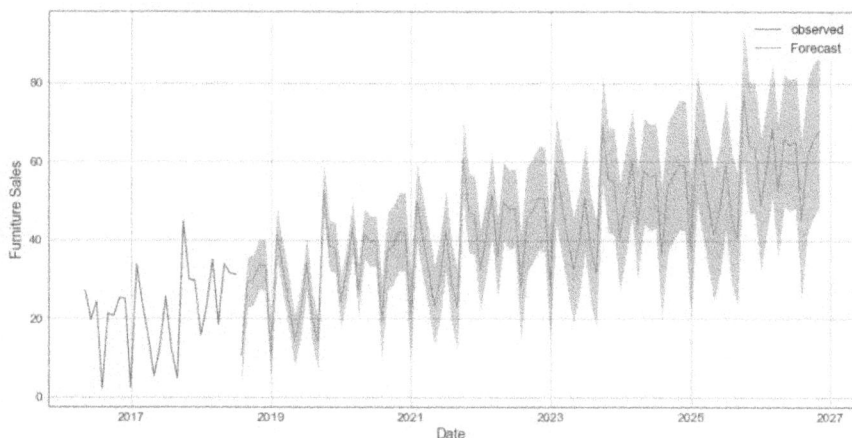

189

To improve this model and obtain more precise forecasts, a good idea would be, for example, to include an external variable such as weather: this would help to understand if there is a correlation between weather and purchases and therefore, would help (possibly) to allocate the marketing budget more effectively.

Analyzing Time-Series with Python

Throughout this chapter, we will see how to analyze time-series using Python with the knowledge we have acquired from the previous chapters of this book.

Before starting to design a trading strategy, it is necessary to analyze the asset on which we will work carefully. Some questions we can ask ourselves are what is its volatility? What was the minimum level of volatility reached historically? How much did the price go back in the worst session? Or the quintessential question; do you tend to have a trend or anti-trend behavior? By analyzing their past behavior, we answer these questions, and this will allow us to focus on the design of our strategy in a much more efficient way.

Here we are going to create a very simple but also an efficient tool in Python that will allow us to characterize the asset we want to analyze. The ultimate purpose will be to obtain information on whether we should approach the design of the strategy for that asset with a trend or anti-trend approach. We can also find that there is no clearly defined bias, which would mean that there is no distinct advantage to design a system based directly on one of these two approaches.

I have divided the complete analysis of a time-series into the following phases. These will be done following the sequential order shown below.

- **Descriptive Statistics:** In this section, we will visually analyze the distribution of returns and obtain a statistical summary of the series.

- **Volatility analysis:** Here, we will analyze the annualized volatility of the series and its historical evolution. In this section, we will also compare the range in bullish and bearish days to determine if there is asymmetry.

- **Hurst Exponent:** We will apply this tool to measure the persistence of the time-series, which will give us information about the trend, anti-trend, or neutral character of the series.

- **Autocorrelation analysis:** Through this analysis, we will measure the relationship between the last closing price and the previous closing prices. This will let us know what impact the previous closures have on future closures.

This analysis also helps us to know if the series has an anti-trend, trend, or neutral behavior.

- *Analysis of the HCLC strategy:* We will program an approach based on a maximum and minimum channel, and we will make a parametric analysis to observe its historical behavior. We will thus check whether there are exploitable patterns in the short, medium, and/or long term using a trend and/or anti-trend approach. Finally, we will register the capital curves of the best trend strategy and the best anti-trend strategy.

- *Summary of results:* We will collect the results obtained in the analysis of the HCLC strategy.

- *Monte Carlo simulation:* Finally, we will make a Monte Carlo simulation of the system that has shown the best behavior (trend or anti-trend).

Used Tools

To design what we have previously called our "tool," we will use a Jupyter Notebook like before. We can access the interactive environment of the Jupyter Notebook through the Anaconda installation. On the official site, there are few versions of Python available for download up to the latest 3.8. Versions below 3 are outdated, so it is not convenient to install one from below 3. Here we are also going to use version 3.7.

We will also access the Yahoo Finance API to download historical end-of-day data. We must consider something important; Yahoo data is not of good quality, and it is usual to find errors in them. We can discover unreal peaks in the quotes, and/or failures in the calculation of splits or dividends. However, here we have used this provider as it also has great advantages; among them, access to the API is free without registration, and it also has a large number of assets with historical lengths. For these reasons, as a starting point, Yahoo seems easy for me to make a quick and convenient exploratory analysis for your convenience.

Descriptive Statistics

Let's start with the creation of the Notebook. The first thing we will do is add the Python libraries that we will use to perform the analysis:

```
import pandas as pd

import numpy as np

import pandas_datareader as dr

import matplotlib.pyplot as plt
```

```
# Library to improve chart display

plt.style.use ('seaborn-darkgrid')
```

I will explain shortly again what each of these libraries is for.

- **_Numpy:_** Acronym of Numerical Python is the library used par excellence for scientific computing and numerical analysis. It includes a large number of functionalities to perform vector and matrix operations.

- **_Pandas:_** Built on Numpy, it is a library used to manipulate and analyze data in Python. It offers data structures and operations to manipulate numerical tables and time-series.

- **_Pandas_datareader:_** It is a sub-library that allows obtaining historical data from different online providers and storing us in a DataFrame. From here, you can see the currently supported providers. This library must be installed manually as it does not come by default in Anaconda. If you need it, here you will find a video tutorial to install it.

Pandas Link: https://pandas-datareader.readthedocs.io/en/latest/remote

Video Link:

- **Matplotlib:** It is a library used to generate graphics from lists or arrays in Python and its mathematical extension Numpy.

Next, we will configure the necessary parameters to download the historical data of the asset that we want to analyze. Here we need to configure the name of the asset, and the range of dates that we will download. We will select the SPDR S&P 500 (SPY) ETF from February 1993 to November 2019. As you can see below, through the Pandas_Datareader library we will download this data and store it in a new DataFrame called "df":

```
# We select the asset to analyze and the dates of study

from datetime import date

Active = 'SPY'

Start Date = '1993-02-01'

End Date = date.today ()
```

Download the series and show the Tail.

df = dr.data.get_data_yahoo (Active, start = Start Date, end = End Date)

df = df[~ df.index.duplicated ()] # We remove duplicates in the index (Problems with cryptocurrencies)

df.tail()

The df.tail() function will allow us to observe the last five rows of this DataFrame:

Date	High	Low	Open	Close	Volume	Adj Close
2019-10-28	303.850006	302.910004	302.940002	303.299988	4.214700e+07	303.299988
2019-10-29	304.230011	302.859985	303.000000	303.209991	4.423800e+07	303.209991
2019-10-30	304.549988	301.989990	303.429993	304.140015	4.958520e+07	304.140015
2019-10-31	304.130005	301.730011	304.130005	303.329987	6.901190e+09	303.329987
2019-11-01	306.190002	304.739990	304.920013	306.140015	7.109800e+07	306.140015

We will now create a new column in our DataFrame that will show the arithmetic percentage return of the adjusted closing price (Adj Close). To do this, we will simply use the Pandas pct_change () function, as shown below:

We create a new column that includes the percentage change of the Closing price under 'R.Arithmetic' column

```python
df ["R. Arithmetic"] = df ["Adj Close"]. pct_change ()

df.tail ()
```

And ready! Our DataFrame will look like this:

Date	High	Low	Open	Close	Volume	Adj Close	R. Aritmetico
2019-10-29	304.230011	302.859985	303.000000	303.209991	44238000.0	303.209991	-0.000297
2019-10-30	304.549988	301.989990	303.429993	304.140015	49585200.0	304.140015	0.003067
2019-10-31	304.130005	301.730011	304.130005	303.329987	69011900.0	303.329987	-0.002663
2019-11-01	306.190002	304.739990	304.920013	306.140015	71098000.0	306.140015	0.009264
2019-11-04	308.000000	307.250000	307.850006	307.494995	29919016.0	307.494995	0.004426

Now we have something to start analyzing. We will begin by obtaining a frequency histogram to know the historical distribution that these returns have had. In this histogram, we will also add the probability density function (FDP) of a normal distribution. This will allow us to know if the historical distribution of returns fits, or not, to a normal distribution. Building this graphic through Python is very simple. We will create it by adding the block of code shown below:

```python
import seaborn as sns

from scipy import stats

from scipy.stats import norm
```

Delete the first row of the DF; It contains a NaN value in the calculation of the Arithmetic R.

df = df.iloc [1:]

We draw the frequency chart.

plt.figure (figsize = (15.8))

sns.set (color_codes = True)

ax = sns.distplot (df ['R. Arithmetic'], bins = 100, kde = False, fit = stats.norm, color = 'green')

We obtain the adjusted parameters of the normal distribution used by SNS

(mu, sigma) = stats.norm.fit (df ['R. Arithmetic'])

Configure the graphic title, legends, and labels.

```
plt.title ('Historical Distribution of Daily Returns', fontsize
= 16)

plt.ylabel ('Frequency')

plt.legend (["Normal distribution. fit ($ \ mu = $ {0:.2g},
$ \ sigma = $ {1:.2f})". format (mu, sigma), "R. Arithmetic
Distribution"])
```

As you can see here, we have first imported Seaborn. Seaborn is a Matplotlib-based data-visualization library that provides a high-level interface for drawing attractive and informative statistical graphs. In the following two lines of code, we import the scipy.stats module that contains a large number of statistical functions, and then import norm. This last function will allow us to create the probability density function (FDP) for the normal distribution. In the following lines of code, we have the necessary instructions to create our frequency histogram (see comments within the code). If we execute this code block, we will obtain a graph like the following:

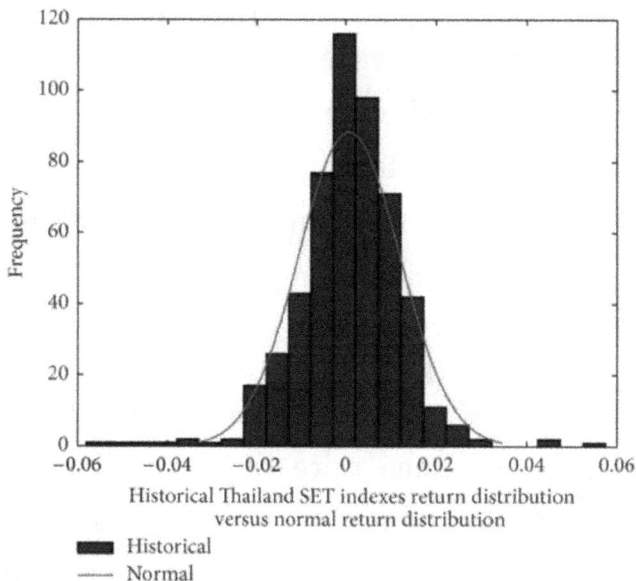

Historical Thailand SET indexes return distribution versus normal return distribution

■ Historical
--- Normal

The histogram consists of an x-axis (horizontal) and a y-axis (vertical). In our case, the blue bars of the histogram shows the frequency (y-axis) with which the arithmetic returns are within a certain range of values of the x-axis. The red line shows the FDP of the theoretical normal distribution. This distribution is computed using the average and standard deviation of our historical return series.

Here we can see that our return distribution has some similarity to a normal distribution; however, it does not fit perfectly. In reality, the series of returns that we find in the markets rarely, if not ever, fit perfectly to a normal distribution. In general, they tend to show extreme values that deviate from their average with a higher probability than expected in a normal distribution. This produces distributions with long tails and, therefore, more likely to suffer tail risk. The coefficient of asymmetry and kurtosis help us to know how much our distribution deviates from a theoretical normal distribution. We will get these two values later.

In the analysis of the financial time-series, by simplification, it is based on the assumption that the price series conform to a normal distribution. The main problem that arises with this theory is that it underestimates the risk of obtaining extreme values. That means if we assume that the distribution is normal and its tails are longer than normal, the chances of getting absolute values will be higher than expected. Therefore, when this occurs, it is necessary to rule out normality and use more efficient tools to measure risk. Among which we have the Value at Risk (VaR) or CVaR was applying the historical calculation method.

Next, we will add the following code block to obtain the descriptive statistics of the returns. Here we have added the statistics that I consider most important. However, if we want, we can customize this section by adding those we consider. Inside the code, we will find comments that tell us what the different instructions are for:

We calculate the Composite Annual Growth Rate and the result of Buy and hold.

Years = df ["R. Arithmetic"]. Count () / 252

CAGR = (df ['Adj Close']. Iloc [-1] / df ['Adj Close']. Iloc [0]) ** (1 / (Years - 1)) - 1

print ('> Compound Annual Growth Rate:', '%. 6s'% (100 * CAGR), '%')

print ('> Buy & Hold:', '%.6s'% (100 * (df [' Adj Close ']. iloc [-1] / df [' Adj Close ']. iloc [0])),' % ')

Calculate the Maximum Drawdown.

Maximum_Previous = df ["Adj Close"]. Cummax ()

drawdowns = 100 * ((df ["Adj Close"] - Maximum_Previous) / Maximum_Previous)

DD = pd.DataFrame ({"Adj Close": df ["Adj Close"],

"Previous Peak": Maximum_Previous,

```
                    "Drawdown": drawdowns})

print ('> Historic MaxDrawdown:', '%.6s'% np.min (DD
['Drawdown']), '%')



# We obtain the average, standard deviation, maximum
and minimum value and number of data analyzed:

print ('> Daily Average:', '%.6s'% (100 * df ["R.
Arithmetic"]. mean ()), '%')

print ('> Daily Typical Deviation:', '%.6s'% (100 * df ["R.
Arithmetic"]. std (ddof = 1)), '%')

print ('> Maximum Daily Loss:', '%.6s'% (100 * df ["R.
Arithmetic"]. min ()), '%')

print ('> Maximum Daily Profit:', '%.6s'% (100 * df ["R.
Arithmetic"]. max ()), '%')

print ('> Days analyzed:', '%.6s'% df ["R. Arithmetic"].
count ())

print ('<---------------------------------------------- ----> ')
```

Coefficient of asymmetry and kurtosis of the distribution.

```python
print ('> Asymmetry Coefficient:', '%.6s'% df ["R. Arithmetic"]. skew ())

print ('> Kurtosis:', '%.6s'% df ["R. Arithmetic"]. kurt ())

print ('<--------------------------------------------- ----> ')
```

Theoretical VaR obtained through the normal distribution at 95% and 99% confidence.

```python
print ('> VaR Gaussian Model NC-95%:', '%.6s'% (100 * norm.ppf (0.05, mu, sigma)), '%')

print ('> VaR Gaussian Model NC-99%:', '%.6s'% (100 * norm.ppf (0.01, mu, sigma)), '%')

print ('> VaR Gaussian Model NC-99.7%:', '%.6s'% (100 * norm.ppf (0.003, mu, sigma)), '%')
```

95% historical VaR and 99% confidence.

```
print ('> VaR Historic Model NC-95%:', '%.6s'% (100 *
np.percentile (df ["R. Arithmetic"], 5)), '%')

print ('> VaR Historic Model NC-99%:', '%.6s'% (100 *
np.percentile (df ["R. Arithmetic"], 1)), '%')

print ('> VaR Historic Model NC-99.7%:', '%.6s'% (100 *
np.percentile (df ["R. Arithmetic"],. 3)), '%')
```

If we execute this block of code we will obtain the following results:

```
>Annual Composite Growth Rate: 9.9425%
>Buy & Hold: 1148.6%
>Maximum Historic Drawdown: -55.18%
>Daily Average: 0.0427 X%
>Typical Daily Deviation: 1.1448 X%
>Maximum Daily Loss: -9,844%
>Maximum Daily Benefit: 14,519%
>Days analyzed: 6742
<------------------------------------------------->
> Asymmetry coefficient: 0.0724
> Kurtosis: 11,089
<------------------------------------------------->
>VaR Gaussian Model NC-95%: -1,840%
>VaR Gaussian Model NC-99%: -2.620%
>VaR Gaussian Model NC-99.7%: -3.102%
>VaR Historic Model NC-95%: -1.807%
>VaR Historic Model NC-99%: -3.106%
>VaR Historic Model NC-99.7%: -4.563%
```

In the first section we have the basic statistics of the series; compound annual growth rate, result of the Buy & Hold "strategy", maximum historical Drawdown, etc. Generally, these are the results that we can use to measure the volatility of the stock market for a certain period.

In the lower section, we have the asymmetry coefficient and the kurtosis coefficient. Both serve to know how returns are distributed in our distribution, and how much they deviate from a normal distribution.

The skewness measures the skewness of the distribution of the returns on the average. This value can be positive, negative, or neutral. In a normal distribution, the asymmetry coefficient is 0 since both tails are symmetrically balanced. In our distribution, this value is positive, which indicates that the right tail is longer than the left. In other words, in the right tail, we find values farther from the average than in the left tail, which in this particular case is positive.

Kurtosis (Kurtosis) defines to what extent the different queues distribution queues a normal distribution. In a normal distribution, kurtosis is 3. In our distribution, this value is higher, which indicates that the tails are longer than expected in a normal distribution. In other words, it demonstrates that our distribution contains values that exceed three standard deviations from the average, which are not exceeded with a probability of 99.7% in a normal distribution. When the kurtosis exceeds normal, we can say that the distribution is leptokurtic.

After analyzing both coefficients, we see that the series does not fully adjust to normal. Now let's see what the limitation of estimating the risk of the series assuming that it follows a normal distribution is.

In the last section of the statistics, we have the VaR calculated in two different ways: In the first case, the Gaussian model is used, which assumes that the returns are normally distributed. In the second case, the model based on historical data is used, in which the empirical distribution is used, and in which no assumptions are made about the type of distribution of the series. In both cases, the worst expected loss is obtained with a confidence level of 95%, 99%, and 99.7%.

Here it is observed that with a confidence level of 95%, both methods obtain a very similar value; however, at 99% and 99.7%, the differences are notable. In both cases, the expected result in the empirical model is worse than in the Gaussian model, which confirms again that the left tail of our distribution is longer than normal.

Volatility Analysis

We will now analyze the historical volatility of our time-series. We are primarily interested in knowing the evolution of volatility throughout the period. The code below helps us to create a graph that shows the evolution of the following three values:

- *Annualized volatility:* Standard deviation of the last 14 days multiplied by the square root of 252.
- *The arithmetic mean on annualized volatility (SMA):* Shows the average of the last 126 values of annualized volatility.
- *The closing price adjusted.*

Within the following block of code, we will find different comments that explain the usefulness of the different instructions:

```
# We register Matplotlib converters to adapt the date -
DateTime- to the Matplotlib drawing method

from                pandas.plotting              import
register_matplotlib_converters

register_matplotlib_converters ()
```

We calculate the historical volatility of 14 days, the annualized volatility and its SMA of 252 days.

```
df ['Historical_ Volatility_14_Days'] = 100 * df ["R. Arithmetic"]. rolling (14).std ()

df ['Volatility_14_Days_Anualized'] = df ['Volatality_Historic_14_Days'] * (252 ** 0.5)

df ['SMA_126_Volatality_Anualized'] = df ['Volatality_14_Days_Anualized']. rolling (126).mean ()
```

We create a graph in which we show the closing price, annualized volatility and its SMA.

```
fig, ax1 = plt.subplots (figsize = (15, 8))

ax2 = ax1.twinx ()

ax1.plot (df ['Volatility_14_ Days_Annualized'], 'orange', linestyle = '-')

ax1.plot (df ['SMA_126_Volatality_Anualized'], 'Green', linestyle = '-')

ax2.plot (df ['Adj Close'], 'black')
```

```
# Assign a name for the graph and for the x and y axes.

plt.title ('Historical Evolution of Price and Volatility',
fontsize = 16)

ax1.set_xlabel ('Date')

ax1.set_ylabel ('Annualized Volatility', color = 'black')

ax2.set_ylabel ('Closing Price', color = 'black')


# Configure the legend, the grid lines, and show the graph.

ax1.legend (loc = 'upper left', frameon = True, borderpad =
1)

ax1.grid (True)

ax2.grid (False)

plt.show ()
```

This code produces the following graphic:

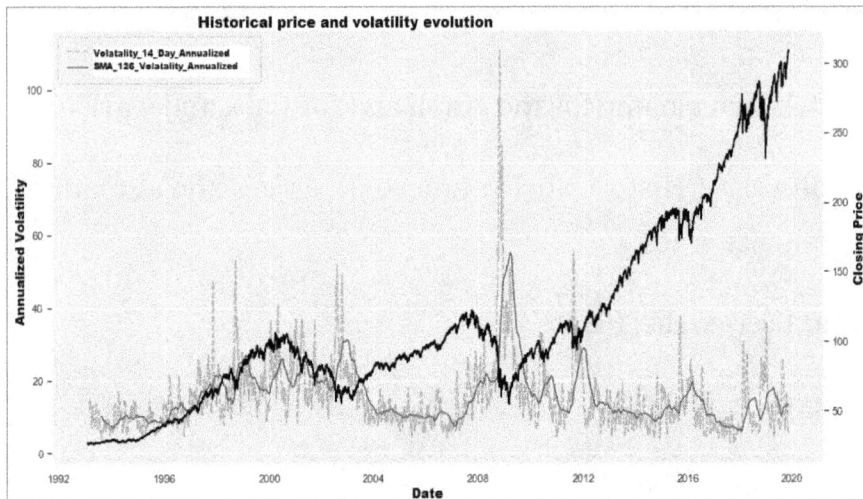

Historical price and volatility evolution

No doubt, a picture is worth a thousand words. Here we can see the historical evolution of annualized volatility, showing the different peaks and valleys produced throughout history, and their relationship with historical price movements.

The next block of instructions will allow us to analyze this volatility further. Here we will begin obtaining historical volatility. Then we will locate what has been the maximum and minimum value historically achieved and record the dates on which both values have been produced. To complete the analysis, we are also interested in knowing if there is any relationship between volatility and the direction of the session. To answer this question, we will calculate the average percentage range of bullish and bearish days independently. And then, we will build a ratio that will compare both values to know the asymmetry of volatility.

Annualized volatility.

VAM = (252 ** 0.5) * (100 * df ["R. Arithmetic"]. Std ())

print ('> Annualized Volatility:', '%. 6s'% VAM, '%')

We obtain the minimum and maximum value of annualized volatility as well as the dates

in which both values are produced.

Date_Minimal_Volatility   =
df.Volatility_14_Days_Annualized
[df.Volatility_14_days_Annualized ==

     df ['Volatility_14_Days_Annualized']. min ()]. index.strftime ('% Y-% m-% d'). tolist ()

Date_Maximum_Volatility   =
df.Volatility_14_Days_Annualized
[df.Volatility_14_Days_Annualized ==

```python
        df ['Volatility_14_Days_Annualized']. max
()]. index.strftime ('% Y-% m-% d'). tolist ()


print ('> The Minimum Annualized Volatility was', '%.6s'%

    (df ['Volatility_14_Days_Annualized']. min ()),

    '%', 'registered on', Date_Minimum_Voltility [0])

print ('> Maximum Annual Volatility was', '%.6s'%

    (df ['Volatility_14_Days_Annualized']. max ()),

    '%', 'registered on', Date_Maximum_Voltility [0])


# We get the average over the percentage range of negative
days.

df ['DaysNegatives'] = np.where (df ['R. Arithmetic'] <0,
100 * (df ['High'] - df ['Low']) / df ['Low'], 0)

df_days_negatives = df.loc [df ['DaysNegatives']! = 0]

DN = df_days_negatives ['DaysNegatives']. Mean ()

print ('> Negative Days Average Range:', '%. 4s'% DN, '%')
```

We obtain the average percentage range of positive days.

df ['Positive Days'] = np.where (df ['R. Arithmetic']> 0, 100 * (df ['High'] - df ['Low']) / df ['Low'], 0)

df_days_positives = df.loc [df ['DaysPositives']! = 0]

DP = df_days_positives ['DaysPositives']. Mean ()

print ('> Medium Range Positive days:', '%. 4s'% DP, '%')

We calculate the ratio of the range between positive and negative days.

print ('> RDP / RDN Ratio', '%. 4s'% (DN / DP), '%')

After executing this group of instructions, we will obtain the following statistics:

```
> Annualized Volatility: 18,159%
> The Minimum Annualized Volatility was 2.4551% recorded on 2017-08-09
> The Maximum Annual Volatility was 110.78% recorded on 2008-10-28
> Average range of negative days: 1.42%
> Mid Range Positive days: 1.21%
> RDP / RDN ratio 1.17%
```

Here we can see that the annualized volatility of the SPY was 18.16% (using all the history). The minimum volatility was recorded in August 2017, with a value of 2.45%, and the maximum volatility in October 2008 is with a value of 110.78%. We also see that the average percentage range of negative days was 1.42%, and that of positive days of 1.21%. The RDP / RDN Ratio indicates that the average volatility of the negative days was specifically 17% higher than the average volatility of the positive days a normal asymmetry in equity indexes.

Conclusion

The analysis of time-series plays a vital role in the study for forecasting future events in different situations. As we have learned throughout the book, there are several ways or methods to calculate what will be the trend of the behavior of a process under the analysis.

Statistical analysis today is greatly facilitated by tools or software that enable faster processing for further investigation and its extensive graphics capabilities, throughout the book, we showed reports and calculations obtained from different angles and situations.

Analysis of occurrences through time-series for studies of processes of sales, changes in behavior regarding consumption, changes in inflation rates time-series will allow us to analyze in a simple way to forecast the future results and depending on the trend analysis technique we will approach with greater or lesser precision the values that will happen.

Any analysis must also consider that the factors that have been occurring in the period to be evaluated will continue to be influenced in the same way in our future scenario. Any sharp or unexpected change in any of the factors may result in non-compliance with the calculated trends.

When choosing which tools to use for statistical analysis, the most important is the programming language, as this is the core component that will help us to run the analysis flawlessly. Python is characterized by being a dynamic, interpreted, agile, fast-growing language with a massive number of libraries that support the entire ecosystem in the Python world.

Dynamic languages such as Python have the advantage of building MVPs (Minimum Viable Product) in record time. This will allow being at the forefront in the market with a product or service ready in times less than the competition.

To achieve all of the above, we must be able to adapt to the market quickly, continually improving and growing gradually.

Python is completely able to provide us with all the resources we need to get the goals set with the time-series forecasting process.

Although a dynamic tool will indeed facilitate your work, the most important reason when choosing Python as a language is a large number of libraries that are needed to get through your analysis.

We have previously mentioned that Python dynamism plays an important role when choosing this tool over others.

A crucial element when choosing what language to use for our project is the support it has at the data provider level. Python is almost omnipresent as a client to access any service in the cloud where we need to access, either through an API, like any other script or access.

Python has libraries that stand out in the world of Data Science, mathematical and statistical analysis, financial risk analysis, machine learning, Macrodata, etc.

Time-series considered as 'the information superhighway' contains a lot of data that makes the analysis and the processing very complicated. Pandas have handy functions that help you process and obtain essential results for the growth of your statistical projects.

Time Series with Python is amazingly useful and beneficial for every business. Best of luck to you as you explore it.

Resources

https://link.springer.com/chapter/10.1007%2F1-84628-184-9_2

https://petrowiki.org/Production_forecasting_decline_curve_analysis

https://www.machinelearningplus.com/time-series/arima-model-time-series-forecasting-python/

https://www.researchgate.net/publication/301232812_From_time_series_to_complex_networks_The_visibility_graph

https://pdfs.semanticscholar.org/15de/cc2d0f9a1ac16765ed144747ccb97fe1a5d0.pdf

https://www.kaggle.com/competitions?sortBy=relevance&group=all&search=electricity&page=1&pageSize=20

https://archive.ics.uci.edu/ml/datasets/ElectricityLoadDiagrams20112014

https://pandas-datareader.readthedocs.io/en/latest/remote_data.html

https://stackoverflow.com/questions/31413934/ howto-get-fit-parameters-from-seaborn-distplot-fit

https://blog.quantinsti.com/ portfolio-optimization-maximum-return-risk-ratio-python/